RECURSIVE STREAMFLOW FORECASTING

UNESCO-IHE LECTURE NOTE SERIES

Recursive Streamflow Forecasting
A State–Space Approach

JÓZSEF SZILÁGYI
Department of Hydraulic and Water Resources Engineering,
Budapest University of Technology and Economics (BUTE), Hungary
School of Natural Resources, University of Nebraska-Lincoln, USA

ANDRÁS SZÖLLÖSI-NAGY
UNESCO-IHE Institute for Water Education, Delft, The Netherlands

CRC Press
Taylor & Francis Group
Boca Raton London New York Leiden

CRC Press is an imprint of the
Taylor & Francis Group, an **informa** business

A BALKEMA BOOK

CRC Press/Balkema is an imprint of the Taylor & Francis Group, an informa business

© 2010 Taylor & Francis Group, London, UK

Typeset by Vikatan Publishing Solutions (P) Ltd., Chennai, India
Printed and bound in Great Britain by Antony Rowe (a CPI Group Company), Chippenham, Wiltshire

Published by: CRC Press/Balkema
    P.O. Box 447, 2300 AK Leiden, The Netherlands
    e-mail: Pub.NL@taylorandfrancis.com
    www.crcpress.com – www.taylorandfrancis.co.uk – www.balkema.nl

Library of Congress Cataloging-in-Publication Data

Szilágyi, József.
   Recursive streamflow forecasting : a state-space approach / József Szilágyi,
   András Szöllösi-Nagy.
      p. cm. -- (UNESCO-IHE lecture note series)
   Includes bibliographical references.
   ISBN 978-0-415-56901-9 (hardcover)
   1. Stream measurements. 2. Streamflow--Forecasting. I. Szöllösi-Nagy, András.
   II. Title. III. Series.

   GB1201.7.S95 2010
   551.48'3--dc22

                              2010017635

ISBN: 978-0-415-56901-9 (Hbk)
ISBN: 978-0-203-84144-0 (eBook)

To Jim Dooge

Contents

Preface

In the light of the current record-breaking floods in Hungary and in Central Europe in the summer of 2002, which caused numerous deaths and property damage in the tens of billions of euros, the value of reliable and accurate streamflow forecasting can be appreciated. By knowing in advance when, where and at what level the river will crest, appropriate flood protection works can be planned and organized, thus reducing possible damage to life and property. Currently there is a wide range of forecasting methods used at different agencies across the world responsible for producing streamflow forecasts. Our work describes in detail the one used by the National Hydrological Forecasting Service in Hungary, a country that, in Central Europe, has the largest proportion of its population (25% of a population of 10 million) working and/or living in flood-plains, that are protected by levees with a total length that is second to none in Europe, including the Netherlands.

In the past there have been publications on streamflow modeling and forecasting, but none of those works concentrated on a single technique in great detail. With the current work, we would like to fill that gap by meticulously going through a detailed derivation of a streamflow modeling technique that (a) is physically based; (b) is formulated with discrete data in mind; (c) accounts for model uncertainties; (d) is adaptive; and (e) is mathematically elegant. Beyond the mathematical and physical background necessary for the derivation of the model, specific examples are shown regarding how the model performs in practical applications. The derivation requires a state-space approach often used in hydrological modeling, but less frequently discussed in detail in the water resources literature and perhaps never discussed in such a thorough, rigorous and step-by-step fashion as here. Without claiming superiority to other streamflow forecasting techniques, a detailed and comprehensive description of the present approach should help water-resources practitioners and graduate students with a shared interest in hydrology to formulate their state-space models for a wide range of applications where linear ordinary or partial differential equations are involved.

CHAPTER 1

Introduction

Traditional handbooks of hydrology (e.g. Shaw, 1983) commonly separate hydrological forecasts into two categories: (a) forecasting of extreme events; and (b) real-time forecasting with a typical objective of describing the physics of the processes to be modeled in partial or full detail. While the former type of forecasts center mainly on issuing flood warnings, the latter provides additional information, such as what is necessary for the optimal operation of water-related infrastructure, on a continuous, operational basis. This way real-time forecasts can incorporate event-based forecasts.

Another classification of real-time, operative forecasting can be drawn based on the lead-time involved. This may present the following categories: (a) general warnings and alerts, based on synoptic meteorological situations; (b) hydrometeorological (long-term) forecasts using measured precipitation and/or snowmelt rates; and (c) hydrological (short-term) forecasts of downstream flood peaks, based on measured, cresting flood levels at upstream sections of the stream network. Undoubtedly, any kind of categorization is subjective and a function of the dynamics of the processes to be forecast. Also, it goes almost without saying that by increasing the lead-time, dynamics play an ever-diminishing role in computations, leading to increased uncertainties in the forecasts which in the extreme become only general outlooks. Therefore, it is very important to quantify the level of reliability with each lead-time of the forecasts. One thing is certain: the forecasts of different lead-times must build upon each other; consequently, any categorization based on lead-time alone is insufficient.

An ideal, real-time, operative forecasting model should satisfy the following prerequisites. It must:

- account for the *physical laws* that govern streamflow;
- explicitly account for forecasting *uncertainties*;
- react, as quickly as possible, to changes that might occur in the watershed due to natural and human causes by modifying its parameters, i.e. must be *adaptive* while having parameters that are sensitive to the above changes;
- be rendered with the most reliable lead-time because models with a short lead-time generally diverge after some critical time, leading to unreliable forecasts;

– specify and produce unbiased forecast errors;
– be able to accommodate any changes in the observation network and the resulting additional information without changes in its structure;
– make data substitution possible through interpolation or finding analogies where there are missing measurements;
– be numerically stable;
– express fast convergence for any numerical scheme in the model;
– have a structure making it possible to include the model in operational systems of water management;
– have recursive algorithms so that the model will run on portable computers with limited memory capacity.

It may be safe to say that, as of today, no universal, operative forecasting model exists, and most probably there will not be any, at least in the near future. At the same time, the generalization of existing models must be accomplished, and the creation of new, ever more general models must be attempted. For the latter, the MIKE SHE (Refsgaard and Storm, 1995) model is a good example. With generalization, we mean that the models should be made as little site-specific as possible. Optimally, a real-time forecasting model accommodates the modular structuring of existing numerical algorithms. Such a modular structure (Bartha and Szöllősi-Nagy, 1982) is illustrated in Fig. 1.1, where each module represents a sub-function within the complete task of hydrological forecasting.

In what follows, we will concentrate on the flow-routing module function, combined with the stochastic–dynamic module, mentioning the

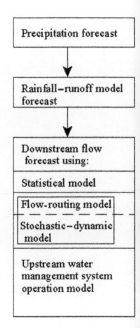

Figure 1.1. Modular structure
of forecasting models.

rainfall–runoff module functions only tangentially. The purely stochastic module will not be mentioned either, because the above two modules replace the former by formulating them to sufficiently account for the physics of the open-channel flow process, while being able to handle uncertainties stemming from stochastic effects. The problem with the application of purely stochastic models lies in the difficulty of interpreting changes in the parameter values, especially when the stream has a short record of measurements. At the same time, a very simple, physically based, deterministic flow-routing model can explain the main tendencies of open-channel flow such that the accuracy of the forecasts deteriorates more slowly with increasing lead-times compared to purely stochastic model forecasts, partly due to more stable parameters in the former. At the same time, physically based, deterministic model forecast errors typically express high autocorrelations, indicating that deterministic models generally cannot fully explain the variance present in the data. Stochastic time-series models, however, are able to extricate this information content of the residuals, paving the way for the combination of the two types of models—deterministic and stochastic—while doing away with the disadvantages of each when used separately. Such a combined deterministic–stochastic model forms the backbone of the unified forecasting system this study reports on.

In order to provide a unified framework for the discussion, comparison, and interpretation of hydrological forecasting approaches, we have to define what is meant by forecasting. This is given by the following definition.

Definition 1: *Let* $\mathbf{y}$ *be the variable (scalar or vector-valued) to be forecasted. Let* $\mathbf{Y}_t$ *be the joint time-series of the present and past values of* $\mathbf{y}$, *such as* $\mathbf{Y}_t = [\mathbf{y}_t, \mathbf{y}_{t-1}, ..., \mathbf{y}_{t-n}]$. *Let* $\mathbf{u}$ *be the variable (scalar or vector-valued) that is in causal relationship with* $\mathbf{y}$, *and let* $\mathbf{U}_t$ *be the joint time-series of the observed present and past, as well as any anticipated future values (denoted by a hat) of* $\mathbf{u}$, *such that* $\mathbf{U}_t = [\widehat{\mathbf{u}}_{t+\tau}, \mathbf{u}_t, \mathbf{u}_{t-1}, ..., \mathbf{u}_{t-n}]$, *and let* $\mathbf{Z}_t = [\mathbf{Y}_t, \mathbf{U}_t]$. *The* $\tau > 0$ *lead-time forecast of the* $\mathbf{y}$ *variable is* $p(\mathbf{y}_{t+\tau} | \mathbf{Z}_t)$, *the conditional probability distribution of* $\mathbf{y}$ *at time* $t + \tau$, *with* $\mathbf{Z}_t$ *as condition.*

Fig. 1.2 displays the forecasted value of a scalar y as a function of the lead-time. The forecast is the conditional expectation of y; the associated standard deviation is an indicator of forecast reliability.

Note that even the observed value (when the lead-time is zero) contains a certain level of uncertainty (i.e. the variance is not zero) due to measurement errors. The above definition is valid for either deterministic or stochastic forecasting methods. In the latter case, a measure of forecast reliability automatically results, but this is not to say that it also means that stochastic forecasting methods are superior to deterministic ones. Clearly, significance levels must be specified for deterministic forecasts as well,

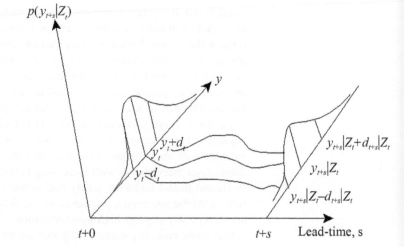

Figure 1.2. Conditional probability, p, forecast of the scalar variable, y, as a function of lead-time. d is standard deviation.

which may be especially critical for decision-makers at times of weighing associated risks and benefits of different actions during extreme events, such as floods. This need requires the augmentation of our deterministic forecasting model with a stochastic model component. When formulating the forecasting model, the objective was to meet as many of the previously laid-out prerequisites of an ideal forecasting model as possible.

During model construction, we were aware that any complex physical system can only be partially described by a purely deterministic model. Consequently, there is always the possibility, or rather the necessity, of including a stochastic model component with the deterministic one, for the purpose of explaining the observed variance in the data missed by the deterministic component. In other words, as long as the time-series of the deterministic model error is autocorrelated, the application of a combined, *deterministic–stochastic* model is justified by not only resulting in forecast confidence intervals but also in improved model forecasts.

To give even a partially comprehensive review of the hydrological fore-casting techniques is beyond the planned framework of this study. Instead, here we just list *some of the earliest works* of real-time, recursive hydro-logical forecasting techniques. These models, almost exclusively, have been formulated in a state–space framework, which first appeared in the 1960s within the field of system/control theory. The state–space frame-work easily allows for applications in automated algorithms of state and parameter *updating*, a task that previously often proved to be very difficult and even impossible in many cases. A system-theoretical description of the hydrological processes in a state–space framework made the applica-tion of filtering techniques possible on digital computers, with the *Kalman filter* being the most famous one. These *digital filters* typically provide fast and effective state and/or parameter updates in a recursive fashion.

Note 1.1:   The application of recursive parameter estimation algorithms first appeared in the hydrologic literature in the early 1970s (Hino, 1974; Szöllősi-Nagy, 1974). Todini and Bouillot (1975) applied recursive parameter estimation in their stochastic rainfall–runoff model using Kalman filtering and Young's technique (1974) of instrumental variables. Szöllősi-Nagy et al. (1977) applied the Kalman filter for parameter estimation in their stochastic hydrologic model. A recursive technique by Bras and Colón (1978) was employed for areal-precipitation estimation, while Kitanidis and Bras (1980) and Georgakakos and Bras (1982) applied an extended version of the Kalman filter for coupled, state and parameter estimation in their nonlinear soil-moisture accounting models. Whitehead (1979) and Moore and Weiss (1980) from the Institute of Hydrology in England researched recursive estimation techniques for simple, conceptual models of hydrology. Cooper and Wood (1982) employed canonical correlations for determining model dimensions in their operative forecasting system. Wood and Szöllősi-Nagy (1978) proposed the application of Bayes-algorithms for adaptive modification of model structure. Recursive state and parameter estimation techniques found their way into water-quality applications (Beck, 1978; Chiu and Isu, 1978; Szöllősi-Nagy, 1979) as well. A good review can be found about the relevant research of the 1960s and 1970s by O'Connell and Clarke (1981). Young's work (1984) on recursive estimation techniques is an excellent textbook on the subject with hydrological examples and references. More recent developments in adaptive real-time flow forecasting are summarized by Young (2002).

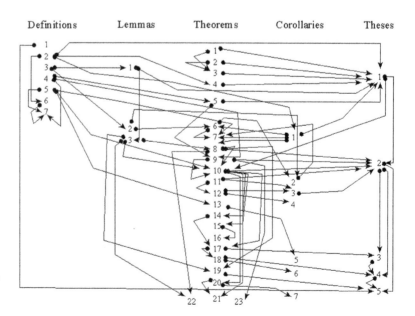

Figure 1.3. Links between definitions, lemmas, theorems, corollaries, and theses in the book.

Finally, some words on the format of the book: Throughout the text, scalar variables are always denoted by *italicised* letters, while bold characters are reserved for vector- or matrix-valued variables. The most important findings are contained in *theorems*, altogether 23 (their proofs included too), which form the backbone of the study. The theorems include 7 *definitions* and the proofs use 3 *lemmas* in all, yielding 7 *corollaries*. The conclusions are summarized in 5 *theses*. The discussion is supplemented with numerous examples, figures, tables, and notes. Each chapter (with the exception of Chapter 9 that discusses some practical aspects of operational forecasting) is closed with a brief summary. Fig. 1.3 depicts the links between the definitions, lemmas, theorems, corollaries, and theses of the study.

CHAPTER 2

Overview of Continuous Flow-routing Techniques

Physically based methods of continuous flow forecasting must necessarily be derived from the *Navier-Stokes equations*. This chapter describes the simplifications which lead to an operative model that meets the prerequisites of the introduction without discussing methods of numerical hydrodynamics that are the subject of Kozák (1977), Brebbia and Ferrante (1983), and Koutitas (1983). Fig. 2.1 summarizes the approaches and models generally used in physically based flow routing.

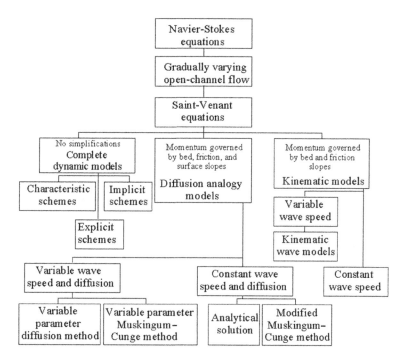

Figure 2.1. Physically based flow-routing approaches (after Jones, 1981).

2.1 BASIC EQUATIONS OF THE ONE-DIMENSIONAL, GRADUALLY VARIED NON-PERMANENT OPEN-CHANNEL FLOW

Flow in river channels is described by the Saint-Venant equations which assume that the flow is mainly one-dimensional and can be characterized by parameters (discharge, cross-sectional area, cross-sectional mean velocity) that are functions only of distance along the river channel (l) and time (t). Derivation of the Saint-Venant equations with ample references and a historical view can be found in Mahmood and Yevyevich (1975), while Szígyártó (1984) provides a semi-empirical derivation of them from the *Navier-Stokes equations*.

The Saint-Venant equations are comprised of the *continuity* or *mass conservation equation*

$$\frac{\partial A}{\partial t} + \frac{\partial Q}{\partial l} = 0 \tag{2.1}$$

where the right-hand-side of Eq. 2.1 is zero only if there is no lateral flow to or from the given stream reach; and the *momentum* or *dynamic equation*

$$S_f = \underset{\substack{\longrightarrow \\ \text{steady}}}{S_0} - \frac{\partial z}{\partial l} - \frac{1}{g}\frac{Q}{A}\frac{\partial\left(\frac{Q}{A}\right)}{\partial l} - \frac{1}{g}\frac{\partial\left(\frac{Q}{A}\right)}{\partial t}. \tag{2.2}$$

$$\underbrace{\phantom{S_f = S_0 - \frac{\partial z}{\partial l} - \frac{1}{g}\frac{Q}{A}}}_{\text{gradually varied steady}} \longrightarrow$$

$$\underbrace{\phantom{S_f = S_0 - \frac{\partial z}{\partial l} - \frac{1}{g}\frac{Q}{A}\frac{\partial}{\partial l}}}_{\text{gradually varied unsteady open-channel flow}} \longrightarrow$$

Here $Q(l, t)$ and $z(l, t)$ are the unknown discharge and stage; A is the cross-sectional area; g is the gravitational acceleration; S_0 is the stream-bottom slope; $S_f = n^2 Q^2 A^{-2} R^{-4/3}$ is the Manning-Strickler friction slope; n is the channel roughness coefficient; and R is the hydraulic radius. Eq. 2.2 in its full form describes a gradually varied, unsteady, open-channel flow.

Eq. 2.2 is of the *hyperbolic* type and can only be solved numerically. The solution, however, requires the simultaneous discharge and/or stage values for the up- and downstream cross-sections of the reach in question at all times (as *boundary conditions*), which means that the Saint-Venant equations could only be used for forecasting purposes if there are already continuous guesses at the downstream discharge to be forecast. This makes the forecasting problem somewhat like a "spatial interpolation" problem between the anticipated simultaneous future discharge values of the two cross-sections for obtaining discharge values along the reach, rather than an "extrapolation" one. Flow forecasting, however, in line with Definition 1, is more like a "spatial extrapolation" problem that specifies the future discharge value at a downstream section of the river as a function of the simultaneous anticipated future discharge value at an upstream location *only,* in addition, of course, to observed discharge values.

Note 2.1:   The anticipated future discharge and/or stage values, as estimated future boundary conditions, at an upstream location can be obtained in the form of flow forecasts using stations even farther upstream or as forecasts of a rainfall–runoff model if no further gauging stations are available.

Any physically based approach, however, must build upon the basic laws of physics that govern open-channel flow. This can be achieved by different simplified forms of the Saint-Venant equations. Eq. 2.3 illustrates these as a function of the degree of simplifications if discharge is expressed from the general friction slope equation as $Q = CR^a\sqrt{S_f}$, $Q_0 = CR^a\sqrt{S_0}$, and Eq. 2.2 is rearranged for Q

$$Q = Q_0\left[1 - \frac{1}{S_0}\frac{\partial z}{\partial l} - \frac{Q}{S_0 Ag}\frac{\partial\left(\frac{Q}{A}\right)}{\partial l} - \frac{1}{S_0 g}\frac{\partial\left(\frac{Q}{A}\right)}{\partial t}\right]^{1/2}.$$ (2.3)

$\underrightarrow{\text{kinematic}}$

$\underrightarrow{\text{diffusion}}$

$\underrightarrow{\text{full dynamic wave}}$

The diffusion wave approach is obtained by neglecting the inertial terms in the full dynamic wave equation; the kinematic wave equation is obtained by further disregarding the water surface slope. The full dynamic wave equation contains the channel roughness coefficient, n, and requires detailed channel geometry information. The former is generally obtained by trial and error; the latter, however, entails the storage and handling of large amount of data, which may be problematic for real-time calculations.

Note 2.2:   Even the full dynamic wave equation provides only an approximate description of gradually varied, unsteady open-channel flow, because it is one-dimensional and the physical content of its parameters is not better founded than those of its simplified versions, since the parameters of the latter can be derived from the former and vice versa (Dooge et al., 1982). Returning to the forecasting paradox when using the Saint-Venant equations, it may be argued that the lower boundary condition could be chosen as sea level or the regulated water levels above a dam on the river. By choosing a large spatial discretization initially with the known water level way downstream, the required lower boundary condition for the given reach could be obtained by successively decreasing the size of the spatial discretization and rerunning the numerical integrations with ever-increasing spatial resolution, finally arriving to the required downstream cross-section of the stream, provided no numerical instabilities are encountered during the process. It remains, however, a question whether this path is worth choosing. A comparative study by Price (1975) concluded that the accuracy of simplified flow-routing techniques generally meet the requirements of practical applications (i.e. even the stringent requirements of real-time forecasting), provided no significant backwater effects are present, and always have superior numerical efficiency over that of the complete dynamic wave equation solution; this latter property being of considerable importance to real-time forecasting. And we have not even mentioned yet that obtaining information on dam operations for determination of the lower boundary conditions would not itself solve the forecasting paradox since dam operations generally depend, among others, on hydrological forecasts which in turn depend, among other things, on dam operations and so on...

2.2 DIFFUSION WAVE EQUATION

When Eq. 2.2 is brought into a dimensionless form, the magnitudes of its terms can be shown to be (Price, 1973)

$$\frac{S_f}{S_0} \sim 0.9 \tag{2.4}$$

$$\frac{1}{S_0}\frac{\partial z}{\partial l} \sim 2.0 \cdot 10^{-2}$$

$$\frac{1}{gaS_0}\frac{\partial}{\partial l}\left(\frac{Q^2}{A}\right) \sim 1.7 \cdot 10^{-3}$$

$$\frac{1}{gaS_0}\frac{\partial Q}{\partial t} \sim \frac{1}{gaS_0}\frac{\partial}{\partial l}\left(\frac{Q^2}{A}\right)$$

which demonstrates that the momentum is affected primarily by the friction slope, S_f, and secondarily by the slope of the water surface. Neglecting the remaining inertial terms, Eq. 2.2 becomes

$$S_f = S_0 - \frac{\partial z}{\partial l} \tag{2.5}$$

as an approximation of the momentum equation.

Henderson (1969) showed that for streams with gently sloping channels, application of Eq. 2.5 is well justified. Eqs. 2.1 and 2.5 can be combined into a single equation by relating the discharge values at the downstream section to that of the upstream location via the hydraulic characteristics of the reach. Differentiating Eq. 2.5 with respect to time, gives

$$\frac{\partial}{\partial t}\left(\frac{\partial z}{\partial l}\right) = \frac{\partial S_0}{\partial t} - \left(\frac{2n^2Q}{A^2R^{4/3}}\frac{\partial Q}{\partial t} - \frac{4n^2Q^2}{3A^2R^{7/3}}\frac{\partial R}{\partial t} - \frac{2n^2Q^2}{A^3R^{4/3}}\frac{\partial A}{\partial t}\right). \tag{2.6}$$

Assuming a rectangular cross-section of width, B, and inserting the cross-sectional area, $A(l,t) = Bz(l,t)$, into the continuity equation

(Eq. 2.1), the following can be written

$$\frac{\partial z}{\partial t} = -\frac{1}{B}\frac{\partial Q}{\partial l}. \tag{2.7}$$

When the mean water depth is much smaller than the stream width, the hydraulic radius, R, can be expressed as, $R(l, t) \simeq z(l, t)$, which upon insertion into Eq. 2.6, together with Eq. 2.7, yields

$$\frac{\partial}{\partial l}\left(\frac{1}{B}\frac{\partial Q}{\partial l}\right) = \frac{2n^2 Q}{A^2 z^{4/3}}\frac{\partial Q}{\partial t} + \frac{4n^2 Q^2}{3A^2 z^{7/3}B}\frac{\partial Q}{\partial l} + \frac{2n^2 Q^2}{A^3 z^{4/3}}\frac{\partial Q}{\partial l} \tag{2.8}$$

This equation, after rearrangement, transforms into a *parabolic nonlinear* partial differential equation (Dooge, 1973)

$$\frac{\partial Q}{\partial t} = D(Q)\frac{\partial^2 Q}{\partial l^2} - C(Q)\frac{\partial Q}{\partial l} \tag{2.9}$$

with

$$D(Q) = \frac{A^2 z^{4/3}}{2n^2 QB} \tag{2.10}$$

and

$$C(Q) = \frac{5}{3}\frac{Q}{A}. \tag{2.11}$$

Eq. 2.9 is known as the *diffusion wave* equation (with zero lateral water flux) because of its similarity with the diffusion equation of turbulent mixing. It is nonlinear because the coefficients, C and D, depend on the unknown variable, Q, posing some problems in the numerical solution similar to the Saint-Venant equations. Hayami (1951) derived the impulse–response of Eq. 2.9 when the coefficients are constants, making the equation linear, and when the lower boundary condition is unspecified, i.e. free. Szöllősi-Nagy (1980) and Ambrus and Szöllősi-Nagy (1984) calculated impulse–responses when the lower boundary condition was specified as well, making use of spatial discretization and a state–space approach, while Dooge et al. (1983) applied Laplace-transforms to obtain the impulse–response. Here we mention that Kontur (1977) solved the diffusion problem in a discrete (in time and space) cascade model framework using a random walk analogy, the first such solution in the field of stochastic hydraulics.

2.3 KINEMATIC WAVE EQUATION

Keeping only the first-order term in Eq. 2.2, gives

$$S_f = S_0 \qquad (2.12)$$

which expresses the balance of the gravitational and dissipation forces. With the Chezy-formula $Q = \varphi(A, l)\sqrt{S_0}$, showing that the discharge is a function (φ) of the cross-sectional area of the water, or simply of the stage, for a rectangular cross-section,

$$\frac{\partial Q}{\partial t} = \frac{\partial Q}{\partial A}\frac{\partial A}{\partial t}$$

can be written. Inserting this identity into Eq. 2.1 results in

$$\frac{\partial Q}{\partial t} + \frac{\partial Q}{\partial A}\frac{\partial Q}{\partial l} = 0,$$

and defining $\partial Q/\partial A = C(Q)$, the *kinematic wave* equation can be written as

$$\frac{\partial Q}{\partial t} + C(Q)\frac{\partial Q}{\partial l} = 0 \qquad (2.13)$$

which is the diffusion wave equation with $D(Q) = 0$ choice. The solution of Eq. 2.13 is

$$Q(l, t) = Q(l - C(Q)t) \qquad (2.14)$$

which shows that the kinematic wave keeps its peak-value as it travels, and if $C(Q) = C$, then it results in a pure translation of the wave without deformation even.

The kinematic wave equation, as the first-order approximation of the Saint-Venant equations, contains very significant simplifications. At the same time, as was shown by Stoker (1953), and Lighthill and Whitham (1955), a significant portion of the flood-wave travels at the speed of the kinematic wave, making methods that assume a single-valued functional relationship between stage and discharge to be quite reliable in general. Notwithstanding, the kinematic wave equation in its original form is unable to explain flood wave attenuation.

2.4 FLOW-ROUTING METHODS

Flow-routing techniques are based on a simplification of the Saint-Venant equations and a postulated relationship with channel storage. Fig. 2.2 lists some of the most popular flow-routing techniques based on a constant

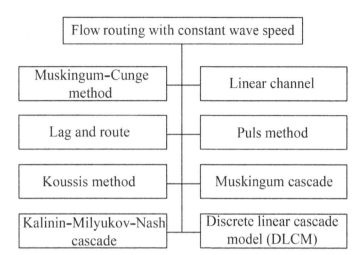

Figure 2.2. Some popular flow-routing techniques applying a constant wave celerity.

wave speed assumption. As Fig. 2.1 shows, they can all be derived from the kinematic wave, Eq. 2.13. The difference between these models is in their spatial discretization schemes and the choice of channel storage function.

2.4.1 Derivation of the storage equation from the Saint-Venant equations

The Saint-Venant equations (Eqs. 2.1 and 2.2) of gradually varied, non-permanent open-channel flow define a system of distributed parameters where the dependent variable is a continuous function of distance along the channel, in addition to time. In practical applications, data is available at specified locations only, requiring the transformation of the partial differential equations into either ordinary differential or algebraic equations, which describe the flow at specified cross-sections of the channel. This entails a lumped parameter system in place of the original distributed parameter one, where now the dependent variable is only a continuous function of time.

For Eq. 2.1, this transformation can be achieved easily by integrating it between the lower (1) and upper (2) boundaries (i.e. cross-sections)

$$\int_1^2 \frac{\partial A}{\partial t} dl = -\int_1^2 \frac{\partial Q}{\partial l} dl$$

which can be written using the Leibniz-rule as

$$\frac{d}{dt} \int_1^2 A(l,t) dl = -Q(l,t) \mid_1^2$$

where the integral on the left-hand-side is the water stored in the reach

$$\int_1^2 A(l,t) dl = S(t)$$

yielding

$$\frac{dS(t)}{dt} = Q_1(t) - Q_2(t) \tag{2.15}$$

where Q_1 is flow into, and Q_2 is flow out of the reach. Eq. 2.15 is the lumped form of the continuity equation and is called the *storage equation*, an integral part to all flow-routing techniques.

Derivation of the lumped version of the momentum equation (Eq. 2.2) is not so simple. Rather, approximate approaches replace Eq. 2.2 with the following relationship

$$S(t) = f[Q_1(t), Q_2(t)] \tag{2.16}$$

which is the other basic equation in flow routing, necessary to make it well defined, since without it the storage equation could not be solved.

Note 2.3:   The continuous operator, f, in Eq. 2.16 can be either differential or algebraic. Examples for the first can be found in Kulandaiswamy (1964), while for the second, linear cascade models discussed below are examples.

2.4.2 *The Kalinin–Milyukov–Nash cascade*

The technique of Kalinin and Milyukov (1957) is based on the concept of the *characteristic reach*. In a characteristic reach, there is a one-to-one relationship between stage and stored water volume. This method assumes that Eq. 2.16 is linear and storage is only a function of the outflow of the reach

$$S(t) = KQ_2(t) \tag{2.17}$$

where K is the mean residence or storage delay time. If there exists a reach for which Eq. 2.17 is valid, then it is a characteristic reach, where the stage-discharge relationship is single-valued, even under unsteady flow conditions. The length (L) of the characteristic reach is given by Kalinin and Milyukov as

$$L = \frac{Q_p}{2S_p \frac{\partial Q_p}{\partial H_p}} \tag{2.18}$$

where Q_p and S_p are discharge and drop in the stage values (between the upper and lower end of the reach) under steady flow conditions; the $\frac{\partial Q_p}{\partial H_p}$ term is the slope of the stage–discharge relationship at H_p. Kalinin and Milyukov showed that the simultaneous changes of the Q_p and $\frac{\partial Q_p}{\partial H_p}$ terms are of about the same magnitude, thus L can be taken as a constant for practical purposes. For a given characteristic reach, Eqs. 2.15 and 2.17 can be combined into a single, linear, ordinary differential equation with constant coefficients

$$K\frac{dQ_2(t)}{dt} + Q_2(t) = Q_1(t). \tag{2.19}$$

The solution of Eq. 2.19, $Q_2(t)$, can be easily computed by the convolution of $Q_1(t)$ with the impulse–response function of Eq. 2.19, which is the outflow response

$$h(t) = \frac{1}{K}e^{-t/K}, \quad t \geqslant 0 \tag{2.20}$$

to an input in the form of the Dirac-delta function, defined as

$$\begin{aligned}
\delta(t) &= 0, \, t \neq 0 \\
\delta(t) &\longrightarrow \infty, \, t = 0
\end{aligned} \tag{2.21}$$

$$\int_{-\infty}^{\infty} \delta(\tau)d\tau = 1.$$

Kalinin and Milyukov further assumed that most river reaches with sufficient length and no lateral in- or outflow can be divided into a series of characteristic reaches of integer number, each with the same storage coefficient. The impulse–response function of a cascade of n serially connected such characteristic reaches can be written as

$$h(t) = \frac{1}{K}\left(\frac{t}{K}\right)^{n-1}\frac{1}{(n-1)!}e^{-t/K}, \quad t \geqslant 0. \tag{2.22}$$

The derivation of the impulse–response through successive convolution can be found in Szöllősi-Nagy (1979). The continuous cascade-model has two parameters (n, the number of characteristic reaches; and K, the mean residence time of the characteristic reach), and gives the flow at the downstream location through convolution of the upstream discharges with Eq. 2.22 as

$$Q_2(t) = \int_{t_0}^{t} h(\tau)Q_1(t - \tau)d\tau. \tag{2.23}$$

Nash (1957) obtained the same impulse–response above for his linear cascade for modeling the relationship between effective precipitation and runoff. For this reason we will call the continuous linear cascade approach the Kalinin–Milyukov–Nash (KMN) cascade.

Note 2.4: Vágás (1970) pointed out that Eq. 2.22 can be interpreted as a *Poisson-distribution* of order ($n - 1$) and parameter $\lambda = t/K$ of the storage delay times

$$P_{n-1}(t) = \frac{1}{(n-1)!}\lambda^{n-1}e^{-\lambda}$$

such that

$$h_{n-1}(t) = \frac{1}{K}P_{n-1}(t). \tag{2.24}$$

For more information on this interpretation, see Bartha and Szöllősi-Nagy (1982); and Diskin (1967) about parameter estimation.

2.4.3 *The Muskingum channel routing technique*

The Muskingum method (McCarthy, 1938) assumes that Eq. 2.16 is linear and storage is a function of either the incoming and outgoing flow of the river reach

$$S(t) = K\left[\varepsilon Q_1(t) + (1-\varepsilon)Q_2(t)\right] \tag{2.25}$$

where ε is a weight, and K is mean residence time. The impulse–response function of the Muskingum model is

$$h(t) = \frac{1}{K(1-\varepsilon)^2}e^{-\frac{t}{K(1-\varepsilon)}} - \frac{\varepsilon}{1-\varepsilon}\delta(t) \tag{2.26}$$

where $\delta(t)$ is the Dirac-delta function. The outflow is again given by the convolution equation (Eq. 2.23)

$$Q_2(t) = \frac{1}{K(1-\varepsilon)^2}\left[Q_1(t_0)e^{-\frac{t}{K(1-\varepsilon)}} + \int_{t_0}^{t}e^{\frac{\tau-t}{K(1-\varepsilon)}}Q_1(\tau)d\tau\right]$$
$$- \frac{\varepsilon}{1-\varepsilon}Q_1(t). \tag{2.27}$$

The last term of the equation is negative; therefore the Muskingum model may give negative outflow discharges when the inflow increases quickly. Cunge (1969) showed that this can be avoided by combining Eq. 2.15 with Eq. 2.25 and applying a certain discretization scheme in the resulting ordinary differential equation. This has become known as the Muskingum–Cunge technique. A detailed discussion on the subject can be found in Mahmood and Yevyevich (1975). Cunge (1969) and later Jones (1981) also pointed out that the Muskingum method can be derived as a numerical algorithm of the linear kinematic wave equation through the application of a proper discretization scheme.

Similarly to the KMN-cascade, the Muskingum method can also be generalized for a cascade of such reaches. Strupczewski and Kundzewicz (1981) showed the results for identical reaches, while Ambrus and

Szöllősi-Nagy (1984) used reaches with varying parameters. It still remains a question whether these generalized models of the Muskingum method result in better forecast accuracy over the KMN-cascade and/or whether the potential increase in accuracy will offset the increased complexity of the mathematical description when used in an operational setting.

In the following we will not discuss the other flow-routing techniques listed in Fig. 2.2, the only exception being the Discrete Linear Cascade Model (DLCM). Indeed, the main focus of this book is to show how the DLCM can be derived, what its properties are and how it can be applied for operational, real-time flow forecasting. Here it suffices to repeat that all of the flow-routing techniques of Fig. 2.2 can be derived from the linear kinematic wave equation, which, after discretization, is capable of describing the observed attenuation of floodwaves.

This chapter provided the initial conditions for the theoretical results of the book. We gave a brief review of continuous flow-routing techniques as simplifications of the Saint-Venant equations, and showed how they can all be viewed as spatially discretized forms of the continuous linear kinematic wave equation. This latter property will be separately proved again for the KMN-cascade. It follows from the discussion above that the distinction between hydrologic and hydraulic flow-routing methods is rather arbitrary and perhaps unnecessary since both approaches share the same physical core. The large data requirement, computational intensity, and the ensuing forecasting paradox of the full dynamic wave approach gives rise to the multitude of simplified flow-routing techniques and their applications in real-time forecasting. One more thing has yet to be accomplished: a temporal discretization adequate for flow forecasting purposes, which will be the subject of the following four chapters.

EXERCISES

2.1. Show that Eq. 2.2 can be brought into the form in Eq. 2.3.
2.2. Derive the nonlinear diffusion wave equation step-by-step for a wide, shallow rectangular channel.
2.3. Prove that Eq. 2.14 satisfies Eq. 2.13.
2.4. Show that the impulse–response function of n serially connected characteristic river reaches (Eq. 2.22) conserves mass.
2.5. Knowing the impulse–response function of the Muskingum model as well as that the arguments $t - \tau$ and τ are interchangeable in the convolution integral (Eq. 2.23), derive Eq. 2.27.

CHAPTER 3

State–Space Description of the Spatially Discretized Linear Kinematic Wave

In this chapter we will show how the kinematic wave (i.e. the solution of the kinematic wave equation, the basis for most flow routing methods) results as a special case of the general state–space approach of linear systems

$$\dot{\mathbf{x}}(t) = \mathbf{F}\mathbf{x}(t) + \mathbf{G}\mathbf{u}(t) \tag{3.1}$$

$$\mathbf{y}(t) = \mathbf{H}\mathbf{x}(t) \tag{3.2}$$

where $\mathbf{u}$ is the input, $\mathbf{y}$ is output, and $\mathbf{x}$ is the *state variable*; and similarly, $\mathbf{G}$ is the *input*, $\mathbf{F}$ is the *state* or *system*, and $\mathbf{H}$ is the *measurement*, or *output matrix*. The dot denotes differentiation with respect to time. Eq. 3.1, called the *state* or *system equation*, and is an ordinary linear differential equation, while Eq. 3.2, the *measurement* or *output equation*, is an algebraic one; together they define a linear, time-invariant dynamic system. Time-invariance here means that the system matrices $\mathbf{\Sigma} = (\mathbf{F}, \mathbf{G}, \mathbf{H})$ are all constant matrices. Appendix I summarizes some of the basic properties of the state–space approach of linear dynamic systems. See Szöllősi-Nagy (1974) for further definitions concerning hydrologic applications of the state–space approach. Here let it suffice to say that the state variable, $\mathbf{x}$, is a mathematical object that links the input of a dynamic system to its output, typically having some physical meaning (such as stored water volumes), although this latter property is not a requirement for application of the general principles of the approach. It should also be mentioned here that the matrix-triplet, $\mathbf{\Sigma}$, always unambiguously characterizes a dynamic system (Kalman, 1961).

3.1 STATE–SPACE FORMULATION OF THE CONTINUOUS, SPATIALLY DISCRETE LINEAR KINEMATIC WAVE

As was shown earlier, the linear kinematic wave is the first-order approximation of the Saint-Venant equations

$$\frac{\partial Q(l,t)}{\partial t} + C\frac{\partial Q(l,t)}{\partial l} = 0. \tag{3.3}$$

Note 3.1:   The kinematic wave formulation was first done by Lighthill and Whitham (1955) for the transformation of flood-waves in long rivers using the theory of small-amplitude waves, which entailed the linearization of the full dynamic equation (Eq. 2.2). It was subsequently used for describing surface runoff (Woolhiser and Liggett, 1967). Kinematic wave theory has now found its way into many scientific disciplines. See Singh (1997) for a comprehensive review of water resources applications of the kinematic wave equation.

The boundary conditions for Eq. 3.3 are

$$Q(0,t) \;=\; Q(l_0,t) \tag{3.4}$$

$$Q(l,t) \;\neq\; \infty, \quad as \;\; l \to \infty, \;\; t > 0$$

which involve an infinitely long river reach in the limit. The same boundary conditions can be applied for a river reach of finite length, without losing generality. In practical hydrological applications, Q is always finite; thus the lower boundary condition can be neglected, i.e. it is called free. Let's divide the river reach into n non-overlapping sections of equal, Δl length (Fig. 3.1).

By applying a *backward difference-scheme* in Eq. 3.3, the following *ordinary differential equation* results for the l_j cross-section

$$\frac{dQ(l_j,t)}{dt} = -C\frac{Q(l_j,t) - Q(l_{j-1},t)}{\Delta l}$$

$$= \frac{C}{\Delta l}Q(l_{j-1},t) - \frac{C}{\Delta l}Q(l_j,t); \quad 1 \leq j \leq n. \tag{3.5}$$

Let's construct the $\mathbf{x}(t)$ state variable to have discharges at cross-sections $l_j, j = 1, 2, ..., n$ as its elements

$$\mathbf{x}(t) = \begin{bmatrix} Q(l_1,t) \\ Q(l_2,t) \\ \vdots \\ Q(l_n,t) \end{bmatrix}$$

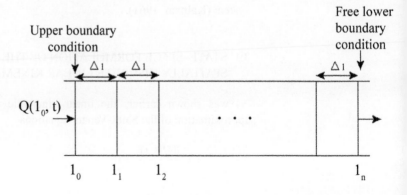

Figure 3.1. Spatial discretization of the linear kinematic wave equation.

and have $\mathbf{u}(t) = u(t) = Q(l_0, t)$, the upper boundary condition (i.e. discharge at the first upstream cross-section). This way Eq. 3.5 becomes

$$\dot{\mathbf{x}}(t) = \frac{C}{\Delta l} \begin{bmatrix} -1 & & & & 0 \\ 1 & -1 & & & \\ & 1 & -1 & & \\ & & \ddots & \ddots & \\ 0 & & & 1 & -1 \end{bmatrix} \mathbf{x}(t) + \begin{bmatrix} \dfrac{C}{\Delta l} \\ 0 \\ 0 \\ \vdots \\ 0 \end{bmatrix} u(t) \qquad (3.6)$$

which in matrix form can be written as

$$\dot{\mathbf{x}}(t) = \mathbf{F}\mathbf{x}(t) + \mathbf{G}u(t), \qquad (3.7)$$

the state equation of a *linear, time-invariant continuous dynamic system*. $\mathbf{F}$ here is a *Toeplitz-band matrix* whose definition can be found in e.g. Rózsa (1974) or Nikolski (2002). Discharge from the last subreach is the discharge of the whole reach; thus the output equation becomes

$$y(t) = [0, 0, ..., 1] \begin{bmatrix} Q(l_1, t) \\ \vdots \\ Q(l_n, t) \end{bmatrix} \qquad (3.8)$$

or

$$y(t) = \mathbf{H}\mathbf{x}(t). \qquad (3.9)$$

The continuous, spatially discrete linear kinematic wave is unambiguously characterized by the matrix-triplet

$$\Sigma_K = (\mathbf{F}, \mathbf{G}, \mathbf{H}). \qquad (3.10)$$

The diagram of the system is illustrated in Fig. 3.2.

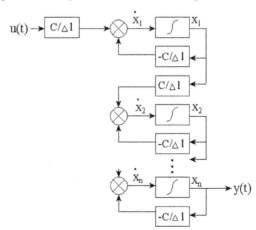

Figure 3.2. System diagram of the continuous, spatially discrete linear kinematic wave.

3.2 IMPULSE RESPONSE OF THE CONTINUOUS, SPATIALLY DISCRETE LINEAR KINEMATIC WAVE

Theorem 1: The impulse response of the continuous, spatially discrete linear kinematic wave, characterized by the matrix-triplet $\Sigma_K = (\mathbf{F}, \mathbf{G}, \mathbf{H})$, is

$$h(t) = \frac{C}{\Delta l}\left(\frac{C}{\Delta l}t\right)^{n-1}\frac{1}{(n-1)!}e^{-\frac{Ct}{\Delta l}}. \tag{3.11}$$

Proof: Being both the input and the output variables scalars, so is the impulse–response function, which can be calculated by Eq. A1.11. The exponential (i.e. the state-transition matrix) of the **F** matrix will be needed. The **F** matrix can be written as

$$\mathbf{F} = \frac{C}{\Delta l}(\mathbf{N}_n - \mathbf{I}_n)$$

where $\mathbf{I}_n$ is $n \times n$ identity matrix, and

$$\mathbf{N}_n = \begin{bmatrix} 0 & & & & \\ 1 & 0 & & & \\ 0 & 1 & \ddots & & \\ \vdots & \ddots & \ddots & \ddots & \\ 0 & \cdots & 0 & 1 & 0 \end{bmatrix}$$

is a nilpotent matrix of order n, the subdiagonal of which (with the unit values) "slips" toward the bottom left corner by each integer increment of its exponent, and the nth power of which is $\mathbf{N}_n^n = 0$. The exponent of the $t\mathbf{F}$ matrix, by definition, can be obtained through Taylor's expansion

$$e^{t\mathbf{F}} = \mathbf{I}_n + t\mathbf{F} + \cdots + \frac{t^n\mathbf{F}^n}{n!} + \cdots = e^{\frac{tC}{\Delta l}\mathbf{N}_n}e^{-\frac{tC}{\Delta l}\mathbf{I}_n}$$

$$= \left[\mathbf{I}_n + \frac{tC/\Delta l}{1!}\mathbf{N}_n + \frac{(tC/\Delta l)^2}{2!}\mathbf{N}_n^2 + \cdots + \frac{(tC/\Delta l)^{n-1}}{(n-1)!}\mathbf{N}_n^{n-1}\right]e^{-\frac{tC}{\Delta l}\mathbf{I}_n}$$

which, in this case, consists of only n terms, since any additional term is zero due to nilpotency. The terms in the expansion are the following matrices:

$$t\frac{C}{\Delta l}\mathbf{N}_n = t\frac{C}{\Delta l}\begin{bmatrix} 0 & & & & \\ 1 & 0 & & & \\ 0 & 1 & \ddots & & \\ \vdots & \ddots & \ddots & \ddots & \\ 0 & \cdots & 0 & 1 & 0 \end{bmatrix}$$

$$\frac{1}{2!}\left(t\frac{C}{\Delta l}\right)^2 \mathbf{N}_n^2 = \frac{1}{2!}\left(t\frac{C}{\Delta l}\right)^2 \begin{bmatrix} 0 \\ 0 & 0 \\ 1 & 0 & \ddots \\ 0 & \ddots & \ddots & \ddots \\ & 0 & 1 & 0 & 0 \end{bmatrix}$$

$$\frac{1}{(n-1)!}\left(t\frac{C}{\Delta l}\right)^{n-1} \mathbf{N}_n^{n-1} = \frac{1}{(n-1)!}\left(t\frac{C}{\Delta l}\right)^{n-1} \begin{bmatrix} 0 \\ 0 & 0 \\ \vdots & 0 & \ddots \\ 0 & \ddots & \ddots & \ddots \\ 1 & 0 & \cdots & 0 & 0 \end{bmatrix}$$

which, when added to the identity matrix, and multiplied by the

$$e^{-\frac{tC}{\Delta l}\mathbf{I}_n} = < e^{-\frac{tC}{\Delta l}} >$$

diagonal matrix, yields the

$$\mathbf{\Phi}(t) = e^{-\frac{tC}{\Delta l}} \begin{bmatrix} 1 & 0 & 0 & \cdots & 0 \\ t\frac{C}{\Delta l} & 1 & 0 & \cdots & 0 \\ \frac{1}{2!}\left(t\frac{C}{\Delta l}\right)^2 & t\frac{C}{\Delta l} & 1 & 0 & \vdots \\ \vdots & \vdots & & \ddots & 0 \\ \frac{1}{(n-1)!}\left(t\frac{C}{\Delta l}\right)^{n-1} & \frac{1}{(n-2)!}\left(t\frac{C}{\Delta l}\right)^{n-2} & & t\frac{C}{\Delta l} & 1 \end{bmatrix}$$

$$(3.12)$$

state-transition matrix. Multiplying the $\mathbf{\Phi}$ lower triangular matrix by the $\mathbf{G}$ column-vector from the right yields the first column of the state-transition matrix times $\frac{C}{\Delta l}$. Multiplying this from the left by vector $\mathbf{H}$, produces the $\mathbf{\Phi}\mathbf{G}$ product's last term, which is Eq. 3.11. This concludes the proof.

It is noted here once again that input to the state–space model is the upstream boundary condition (i.e. inflow discharge series to the reach) of the kinematic wave. There is no need to specify any downstream boundary condition for the calculation of the impulse response. The downstream boundary condition (i.e. outflow discharge series from the reach) is calculated by convolution of the impulse response and the upstream boundary condition. This way stream flow at the downstream cross-section can be calculated without specifying the lower boundary condition, required for the full dynamic wave.

Note 3.2:   Calculation of the state-transition matrix is generally not an easy task. The mathematical literature offers numerous techniques (see Moler and van Loan [1978] for a critical review), starting with the Cayley-Hamilton theorem to the full spectral decomposition of the state matrix, **F**. A general solution, however, does not exist: the procedure to follow depends strongly on the structure of the **F** matrix. For the kinematic wave case, however, the calculation of the state-transition matrix is very simple.

A simple watershed model in state-space

Example 3.1:   The illustration below depicts a simple hydrological system (e.g. a simplified watershed with two subcatchments) where $u_1(t)$ and $u_2(t)$ are the rainfall inputs measured at different locations; the states are defined as the surface storages $x_1(t)$, $x_2(t)$, and $x_3(t)$ and the groundwater storage as $x_4(t)$, respectively. The constants in each case are: k's for surface water flow, and l_1 and l_2 for infiltration. The expression $l_3[x_4(t) - x_3(t)]$ signifies the exchange between the groundwater and the stream. The outputs are $y_1(t)$ and $y_2(t)$, the streamflow output and the contribution of groundwater to streamflow, respectively.

The continuity equations for this problem are

$$\dot{x}_1(t) = -(k_1 + l_1)x_1(t) + u_1(t)$$
$$\dot{x}_2(t) = -(k_2 + l_2)x_2(t) + u_2(t)$$
$$\dot{x}_3(t) = k_1x_1(t) + k_2x_2(t) + l_3[x_4(t) - x_3(t)] - k_3x_3(t)$$
$$\dot{x}_4(t) = l_1x_1(t) + l_2x_2(t) - l_3[x_4(t) - x_3(t)].$$

In vector-matrix form we have the following time-invariant continuous state equation with the initial condition $\mathbf{x}(0) = \mathbf{C}$, a constant vector,

$$\dot{\mathbf{x}}(t) = \mathbf{F}\mathbf{x}(t) + \mathbf{G}\mathbf{u}(t)$$

where

$$\mathbf{F} = \begin{bmatrix} -(k_1 + l_1) & 0 & 0 & 0 \\ 0 & -(k_2 + l_2) & 0 & 0 \\ k_1 & k_2 & -(k_3 + l_3) & l_3 \\ l_1 & l_2 & l_3 & -l_3 \end{bmatrix}$$

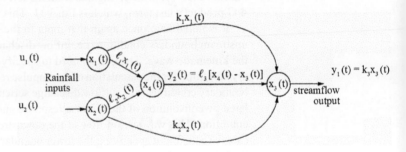

Figure 3.3. A simplified catchment model.

$$G = \begin{bmatrix} 1 & 0 \\ 0 & 1 \\ 0 & 0 \\ 0 & 0 \end{bmatrix}.$$

The output equation becomes

$$y(t) = Hx(t)$$

where

$$H = \begin{bmatrix} 0 & 0 & k_3 & 0 \\ 0 & 0 & -l_3 & l_3 \end{bmatrix}.$$

In this example the states have been defined as storages, i.e. a concrete physical meaning can be attached to them. The following example illustrates that it is not necessary, in general.

The Kulandaiswamy model

Example 3.2: As discussed in detail by Duong et al. (1975), direct runoff may be considered as the result of the transformation of rainfall excess by the basin. The physical process of this transformation is very complex, depending mainly upon the storage effects in the basin. (The reader interested in the details and interconnections between the processes involved is referred to Dooge's (1973) comprehensive review.) To take these effects into account, Kulandaiswamy (1964) derived the following general expression

$$S(t) = \sum_{n=0}^{N} a_n(q,u)\frac{d^n q}{dt^n} + \sum_{m=0}^{M} b_m(q,u)\frac{d^m u}{dt^m},$$

where S is the storage, t is time, and $a_n(q,u)$ and $b_m(q,u)$ are parametric functions of the direct runoff, q, and the excess rainfall, u. To apply the above storage relations to the study of the rainfall–runoff processes in a particular watershed, the values of N, M, and the form of $a_n(\cdot)$ and $b_m(\cdot)$, respectively, must be determined. Unfortunately, it is not always feasible in practice. Therefore Prasad (1967) suggested the application of a simplified storage equation in the form

$$S(t) = K_1 q^N(t) + K_2 \frac{dq(t)}{dt},$$

where K_1, K_2, and N are unknown parameters to be estimated. In his study, Prasad (1967) assumed that these parameters are constant for a particular hydrograph. Employing the continuity equation, the following differential equation is obtained for the rainfall–runoff process

$$K_2 \frac{d^2q}{dt^2} + K_1 Nq^{N-1} \frac{dq}{dt} + q = u.$$

This can be written as

$$\frac{d^2q}{dt^2} = -\left(\frac{1}{K_2}\right) K_1 Nq^{N-1} \frac{dq}{dt} - \left(\frac{1}{K_2}\right) q + \left(\frac{1}{K_2}\right) u.$$

By defining the following set of state variables

$$
\begin{aligned}
x_1(t) &= q(t) \\
x_2(t) &= \dot{q}(t) \\
x_3(t) &= K_1 \\
x_4(t) &= K_2^{-1} \\
x_5(t) &= N \cdot
\end{aligned}
$$

and assuming that the model coefficients are time-invariant, the Prasad model becomes

$$
\begin{bmatrix}
\dot{x}_1(t) \\
\dot{x}_2(t) \\
\dot{x}_3(t) \\
\dot{x}_4(t) \\
\dot{x}_5(t)
\end{bmatrix}
=
\begin{bmatrix}
x_2(t) \\
-x_3(t)x_4(t)x_5(t)x_1^{x_5(t)-1}(t)x_2(t) + x_4(t)[u(t) - x_1(t)] \\
0 \\
0 \\
0
\end{bmatrix}
$$

or in abbreviated notation

$$\dot{\mathbf{x}}(t) = f_t[\mathbf{x}(t), u(t)]$$

which is a time-invariant nonlinear state equation. As for the output equation, it can immediately be seen that by choosing the output process, $q(t)$, as being a state variable itself, it is in the form of

$$
y(t) = [1 \quad 0 \quad 0 \quad 0 \quad 0]
\begin{bmatrix}
x_1(t) \\
x_2(t) \\
x_3(t) \\
x_4(t) \\
x_5(t)
\end{bmatrix}
$$

or

$$y(t) = h_t[\mathbf{x}(t)].$$

In fact, the output equation for the Prasad model is a linear one and the output process is scalar. The conclusions of this example are: (a) it is not necessary for a nonlinear output equation to be attached to a nonlinear state equation; and (b) variables with no direct physical meaning can also be chosen as state variables. Maidment (1975) linearized the Kulandaiswamy model in a state–space fashion.

Wet and dry days
as a Markov chain

Example 3.3:   Gabriel and Neumann (1962) found that a two-state Markov chain yields a good description of the consecutive occurrences of wet and dry days. If p_1 denotes the probability that a dry day is followed by a wet one then $1 - p_1$ denotes the probability of the event that a dry day is followed by another dry day. Similarly, if p_2 denotes the probability that a wet day is followed by a dry one then $1 - p_2$ yields the probability of a wet-to-wet transition. This way the following transition–probability matrix can be constructed

$$\mathbf{\Phi} = \begin{bmatrix} 1 - p_1 & p_2 \\ p_1 & 1 - p_2 \end{bmatrix}$$

which here will play the role of the state-transition matrix and is assumed to be time-invariant. Of course, $0 \le p_1 \le 1$ and $0 \le p_2 \le 1$. Let the vector $\mathbf{x}(t+1) = [x_0(t+1), x_1(t+1)]^T$ denote the probability of finding the system in stage 0 (dry day) or in stage 1 (wet day) at time $t + 1$. Let the initial condition for this vector to be $\mathbf{x}(0) = [x_0(0), x_1(0)]^T$. First, consider the event of being in stage 0 at time $t + 1$. This event can occur in two mutually exclusive ways: (a) from stage 0 at time t no transition out of it occurs at time $t + 1$, having a probability of $x_0(t)(1 - p_1)$; and (b) from stage 1 at time t a transition to stage 0 takes place at time $t + 1$ with an associated probability of $x_1(t)p_2$. The probability of being in stage 1 at time $t + 1$ can be obtained similarly. The probabilities at time $t + 1$ are given by the recurrence relations

$$x_0(t+1) = x_0(t)(1 - p_1) + x_1(t)p_2$$
$$x_1(t+1) = x_0(t)p_1 + x_1(t)(1 - p_2)$$

or in vector-matrix form

$$\mathbf{x}(t+1) = \mathbf{\Phi}\mathbf{x}(t)$$

which is an unforced or free state equation with a solution

$$\mathbf{x}(t) = \mathbf{\Phi}^t\mathbf{x}(0).$$

The related output equation has the form

$$\mathbf{y}(t) = \mathbf{H}\mathbf{x}(t)$$

where $\mathbf{H} = \mathbf{I}$ is the identity matrix, i.e. the states themselves are the output variables. The tth power of the state-transition matrix can be easily calculated with the help of the Cayley-Hamilton theorem as

$$\Phi^t = \frac{1}{p_1 + p_2} \begin{bmatrix} p_2 & p_2 \\ p_1 & p_1 \end{bmatrix} + \frac{(1 - p_1 - p_2)^t}{p_1 + p_2} \begin{bmatrix} p_1 & -p_2 \\ -p_1 & p_2 \end{bmatrix}$$

provided $p_1 + p_2 \neq 0$. Since $\lambda_1 = 1$ and $\lambda_2 = 1 - p_1 - p_2$ are eigenvalues of Φ, and taking into consideration the fact that $x_0(0) = 1 - x_1(0)$, the final results for the probabilities are

$$x_0(t) = \frac{p_2}{p_1 + p_2} + (1 - p_1 - p_2)^t \left[x_0(0) - \frac{p_2}{p_1 + p_2} \right]$$

$$x_1(t) = \frac{p_2}{p_1 + p_2} + (1 - p_1 - p_2)^t \left[x_1(0) - \frac{p_1}{p_1 + p_2} \right]. \tag{i}$$

One question that arises is whether after a sufficiently long period of time the system settles down to a condition of statistical equilibrium in which the state probabilities are independent of the initial condition. If this is so then there is an equilibrium probability $\mathbf{x}^* = [x_0^*, x_1^*]^T$, which, on letting $t \to \infty$, will satisfy

$$\mathbf{x}^* = \Phi \mathbf{x}^*$$

or

$$(\mathbf{I} - \Phi)\mathbf{x}^* = 0$$

which will have nonzero solutions if the determinant $|\mathbf{I} - \Phi|$ vanishes. With this and with the condition $x_0^* + x_1^* = 1$ in mind, the equilibrium probabilities are obtained as

$$x_0^* = \frac{p_2}{p_1 + p_2}$$

$$x_1^* = \frac{p_1}{p_1 + p_2}$$

which are indeed independent of the initial condition $\mathbf{x}(0)$. The equilibrium probabilities might in fact be obtained by taking the limit of $t \to \infty$ in Eq. (i) since $|\lambda_2| < 1$. Finally, for the sake of completeness, consider the degenerate cases. When $p_1 = p_2 = 0$ then

$$\mathbf{x}(t + 1) = \mathbf{x}(t) = \mathbf{x}(0)$$

i.e. the system remains forever in its initial state. On the other hand, if $p_1 = p_2 = 1$ then

$$x_0(t + 1) = x_1(t) = x_0(t - 1) = \cdots$$

$$x_1(t + 1) = x_0(t) = x_1(t - 1) = \cdots$$

which means that the system oscillates deterministically between its two stages, and once the initial state is specified, the behavior of the system is non-random.

This chapter described the state–space derivation of the continuous, linear kinematic wave. The state-transition matrix, i.e. the matrix exponential of the state matrix, could be calculated analytically, which led to specifying the impulse response of the model.

EXERCISES

3.1. Can you guess what the elements of the state-transition matrix in Eq. 3.12 represent in each row?

3.2. From Appendix I, it follows that the impulse response function of the continuous, spatially discrete linear kinematic wave can be written as $h(t) = \mathbf{H}\boldsymbol{\Phi}(t)\mathbf{G}$. Show that it is true for arbitrary n.

3.3. Plot the impulse response functions for $n = 1...5$ with $k = c/\Delta l = 0.5$.

CHAPTER 4

State–Space Description of the Continuous Kalinin–Milyukov–Nash (KMN) Cascade

The basic assumptions behind the continuous KMN-cascade have been discussed in 2.4.2. Using the state–space approach, the model will be redefined here in the hope that it will illuminate not only the compactness but also the elegance of the state–space framework.

Let's start with a scalar case, and consider one single linear storage element with $u(t)$ and $y(t)$ as in- and outflows, respectively. Change in stored water volume, $x(t)$, is described by the continuity equation (Eq. 2.15)

$$\dot{x}(t) = -y(t) + u(t).$$

The dynamic equation now is

$$x(t) = Ky(t)$$

which, when inserted into the above continuity equation, yields the state equation (see Eq. 3.1) of the linear storage element

$$\dot{x}(t) = -\frac{1}{K}x(t) + u(t). \tag{4.1}$$

The corresponding output equation (see Eq. 3.2) is

$$y(t) = \frac{1}{K}x(t) \tag{4.2}$$

since the outflow is directly proportional to the stored water volume.

4.1 STATE EQUATION OF THE CONTINUOUS KMN-CASCADE

The structure of the linear, time-invariant, continuous KMN-cascade is illustrated in Fig. 4.1. The cascade is made up of serially connected storage elements. The output of a storage element is input to the next element in the series, while the output of the last storage element is the output of the whole system.

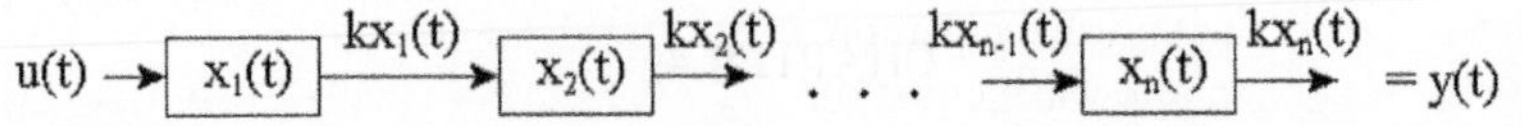

Figure 4.1. Structure of the continuous KMN-cascade.

$$u(t) \rightarrow \boxed{x_1(t)} \xrightarrow{kx_1(t)} \boxed{x_2(t)} \xrightarrow{kx_2(t)} \cdots \xrightarrow{kx_{n-1}(t)} \boxed{x_n(t)} \xrightarrow{kx_n(t)} = y(t)$$

For simplicity, let's define $k = 1/K$. The continuity equation of n storage elements then becomes

$$\begin{bmatrix} \dot{x}_1(t) \\ \dot{x}_2(t) \\ \dot{x}_3(t) \\ \vdots \\ \dot{x}_n(t) \end{bmatrix} = \begin{bmatrix} -k & & & & 0 \\ k & -k & & & \\ & k & -k & & \\ & & \ddots & \ddots & \\ 0 & & & k & -k \end{bmatrix} \begin{bmatrix} x_1(t) \\ x_2(t) \\ x_3(t) \\ \vdots \\ x_n(t) \end{bmatrix} + \begin{bmatrix} 1 \\ 0 \\ 0 \\ \vdots \\ 0 \end{bmatrix} u(t) \quad (4.3)$$

or in matrix notation

$$\dot{\mathbf{x}}(t) = \mathbf{F}\mathbf{x}(t) + \mathbf{G}u(t) \tag{4.4}$$

where $\mathbf{F}$ is $n \times n$ Toeplitz-band state matrix, and $\mathbf{G}$ is $n \times 1$ input matrix/vector (with p-dimensional vector-valued input, it is an $n \times p$ matrix). The corresponding output equation is

$$y(t) = [0, 0, \cdots, k] \begin{bmatrix} x_1(t) \\ x_2(t) \\ x_3(t) \\ \vdots \\ x_n(t) \end{bmatrix} \tag{4.5}$$

or, using matrix notation

$$y(t) = \mathbf{H}\mathbf{x}(t) \tag{4.6}$$

where $\mathbf{H}$ now is a $1 \times n$ matrix, i.e. an n-dimensional row vector.

Eqs. 4.4 and 4.6 define a linear, time-invariant, continuous dynamic system, which is unambiguously characterized by the

$$\Sigma_{KMN} = (\mathbf{F}, \mathbf{G}, \mathbf{H}) \tag{4.7}$$

matrix-triplet. Fig. 4.2 displays the system diagram, which shows striking structural similarity with Fig. 3.2 of the continuous, spatially discrete linear kinematic wave. Below it is shown why.

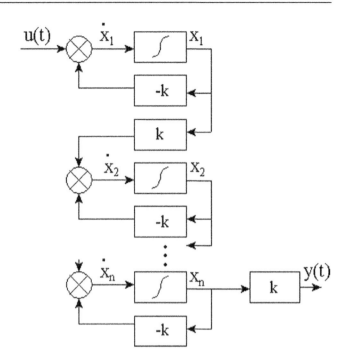

Figure 4.2. System diagram of the continuous KMN-cascade.

4.2 IMPULSE–RESPONSE OF THE CONTINUOUS KMN-CASCADE AND ITS EQUIVALENCE WITH THE CONTINUOUS, SPATIALLY DISCRETE, LINEAR KINEMATIC WAVE

The classical derivation of the impulse–response of the continuous KMN-cascade has already been discussed in 2.4.2. The following theorem therefore does not convey new information. However, it illustrates how elegantly and quickly the state–space formalism leads to results.

Theorem 2: The impulse–response of the continuous KMN-cascade, characterized by $\Sigma_{KMN} = (\mathbf{F}, \mathbf{G}, \mathbf{H})$, is

$$h(t) = k(tk)^{n-1} \frac{1}{(n-1)!} e^{-tk}. \tag{4.8}$$

Proof: The $\mathbf{F}$ system matrix can again be decomposed into the difference of a nilpotent and an identity matrix

$$\mathbf{F} = k(\mathbf{N}_n - \mathbf{I}_n)$$

by which the state-transition matrix can be obtained, as before,

$$
\Phi(t) = e^{t\mathbf{F}} = \begin{bmatrix}
e^{-tk} & 0 & 0 & \cdots & 0 \\
tke^{-tk} & e^{-tk} & 0 & \cdots & 0 \\
\dfrac{(tk)^2}{2!}e^{-tk} & tke^{-tk} & e^{-tk} & 0 & \vdots \\
\vdots & \vdots & \ddots & \ddots & 0 \\
\dfrac{(tk)^{n-1}}{(n-1)!}e^{-tk} & \dfrac{(tk)^{n-2}}{(n-2)!}e^{-tk} & \cdots & tke^{-tk} & e^{-tk}
\end{bmatrix}.
$$

$$(4.9)$$

Multiplying the $\Phi(t)$ matrix by column-vector, $\mathbf{G}$, from the right results in the first column of the state-transition matrix, which, when multiplied by the row-vector, $\mathbf{H}$, from the left, yields the last element of it, times k, i.e.

$$
\frac{k(tk)^{n-1}}{(n-1)!}e^{-tk}
$$

which is Eq. 2.22 with $k = 1/K$. This concludes the proof.

Note 4.1:   Elements in the first column of the state-transition matrix of Eq. 4.9, times k, are the impulse–responses of continuous KMN-cascades of increasing order.

Note 4.2:   There is a notable *duality* between the state–space models of the linear kinematic wave and the KMN-cascade. The $\mathbf{F}$ system matrix is of identical structure in both cases. The $\mathbf{G}$ and $\mathbf{H}$ vectors differ. However, only the first and last elements, respectively, are different from zero in either case.

Even more interesting than duality, is the fact that the linear kinematic wave and the cascade model are the same from a system theoretical point of view. This claim is formulated by the following:

Theorem 3:   The continuous, spatially discrete linear kinematic wave, given by Σ_K, and the continuous KMN-cascade, characterized by Σ_{KMN}, are equivalent.

Proof: Two dynamic systems are equivalent (Desoer, 1970) if their impulse–responses are the same. Eqs. 4.8 and 3.11 are indeed equal with the $k = \frac{C}{\Delta l}$ substitution. This concludes the proof.

Note 4.3:   Equivalence of the two models must show up in the dimensions too. The flow velocity, C, has a unit of distance over time. Δl has a unit of distance; thus the coefficient, $K = \frac{1}{k}$, must have a unit of time, which is true.

4.3 CONTINUITY, STEADY STATE, AND TRANSITIVITY OF THE KMN-CASCADE

Let's now investigate the characteristic properties of the continuous KMN-cascade.

Definition 2: A scalar input/scalar output (SISO) continuous, linear system is *conservative* if

$$\int_0^\infty h(\tau)d\tau = 1 \tag{4.10}$$

where $h(t)$ is the impulse–response function of the system.

Note 4.4: The above definition states that the system is free of any net sources or sinks (Diskin and Boneh, 1972). This is because $h(t)$ is the output of an initially relaxed linear system (i.e. $\mathbf{x}(0) = \mathbf{0}$) to the Dirac-delta function, $\delta(t)$, as input. Since $\int_0^\infty \delta(\tau)d\tau = 1$, and $\delta(t) = 0$ for $t > 0$, the system becomes relaxed again as $t \longrightarrow \infty$. Thus for large enough times ($t \longrightarrow \infty$), total outflow $\int_0^\infty h(\tau)d\tau$ must equal total inflow, if mass is conserved, which is unity by definition of the Dirac-delta function.

Theorem 4: Continuity applies for the Σ_{KMN} continuous KMN-cascade.

Proof: According to Definition 2, the continuous KMN-cascade is conservative (i.e. continuity applies to it) if the area under its impulse–response function is unity:

$$\int_0^\infty h(\tau)d\tau = \int_0^\infty k\frac{(\tau k)^{n-1}}{(n-1)!}e^{-\tau k}d\tau = \frac{k^n}{(n-1)!}\int_0^\infty \tau^{n-1}e^{-\tau k}d\tau. \tag{i}$$

With the $\tau k = t$ substitution, Eq. (i) transforms into

$$\frac{k^n}{(n-1)!}\int_0^\infty \left(\frac{t}{k}\right)^{n-1}e^{-t}\frac{1}{k}dt = \frac{1}{(n-1)!}\int_0^\infty t^{n-1}e^{-t}dt = \frac{\Gamma(n)}{\Gamma(n)} = 1$$

where the definition of the gamma function and the identity, $(n-1)! = \Gamma(n)$, were used. This concludes the proof.

Definition 3: A continuous dynamic system is in a steady state, if

$$\dot{\mathbf{x}}(t) = \mathbf{0} \tag{4.11}$$

(see Csáki, 1973).

Later, the steady state of the KMN-cascade will be needed. The following is related to the issue:

Lemma 1: In a steady state, each storage element of the Σ_{KMN} continuous KMN-cascade has the same amount of water

$$x_i = \frac{1}{k}u_s, \quad i = 1, 2, \cdots, n \qquad (4.12)$$

independent of the total number of storage elements in the cascade. u_s is constant inflow. The total water stored in the cascade is

$$S = Knu_s. \qquad (4.13)$$

Proof: According to Eq. 4.11, in a steady state

$$\mathbf{Fx} + \mathbf{G}u_s = 0$$

from which the steady state system variable becomes

$$\mathbf{x}_s = -\mathbf{F}^{-1}\mathbf{G}u_s. \qquad (i)$$

For obtaining the inverse of $\mathbf{F}$, one can start from the identity

$$\mathbf{F} = k(\mathbf{N} - \mathbf{I}).$$

With this, the inverse of $\mathbf{F}$ can be written as the following matrix-polynomial

$$\mathbf{F}^{-1} = -\frac{1}{k}(\mathbf{I} - \mathbf{N})^{-1}.$$

Similarly to the scalar polynomial identity

$$(1 - z)(1 + z + z^2 + \cdots + z^{n-1}) = 1 - z^n$$

the following can be written

$$(\mathbf{I} - \mathbf{N})(\mathbf{I} + \mathbf{N} + \mathbf{N}^2 + \cdots + \mathbf{N}^{n-1}) = \mathbf{I} - \mathbf{N}^n = \mathbf{I}$$

due to nilpotency. Thus

$$\mathbf{F}^{-1} = -\frac{1}{k}(\mathbf{I} - \mathbf{N})^{-1} = -\frac{1}{k}(\mathbf{I} + \mathbf{N} + \mathbf{N}^2 + \cdots + \mathbf{N}^{n-1})$$

or

$$\mathbf{F}^{-1} = -\frac{1}{k}\begin{bmatrix} 1 & 0 & \cdots & 0 \\ 1 & 1 & \ddots & \vdots \\ \vdots & \vdots & \ddots & 0 \\ 1 & 1 & \cdots & 1 \end{bmatrix}.$$

The steady state system variable in Eq. (i) this way can be expressed as

$$\mathbf{x}_s = -\mathbf{F}^{-1}\mathbf{G}u_s = \frac{1}{k}\begin{bmatrix} 1 & 0 & \cdots & 0 \\ 1 & 1 & \ddots & \vdots \\ \vdots & \vdots & \ddots & 0 \\ 1 & 1 & \cdots & 1 \end{bmatrix}\begin{bmatrix} 1 \\ 0 \\ \vdots \\ 0 \end{bmatrix}u_s = \frac{1}{k}\begin{bmatrix} 1 \\ 1 \\ \vdots \\ 1 \end{bmatrix}u_s.$$

It can be seen that in each storage element, the stored water volume is u_s/k. The steady state outflow is

$$y_s = \mathbf{H}\mathbf{x}_s = k\frac{1}{k}u_s = u_s$$

and the total volume of water stored in the n-order cascade is

$$S = n\frac{1}{k}u_s = nKu_s$$

which concludes the proof.

Corollary 1: If for a SISO linear, time-invariant system the bounded outputs ($|y(t)| < \infty$, $\forall t$) equal the bounded inputs in a steady state, then the system is conservative.

Proof: Once the linear, time-invariant system reaches a steady state at t_0, the system variable is constant (see Definition 3), $\mathbf{x}(t) = \mathbf{x}_0$, until the input, $u(t) = u_s = const.$ for $t > t_0$. The steady state output is now equal to the constant input, u_s. Applying Eq. A1.5, the output can be written as

$$u_s = \mathbf{H}\boldsymbol{\Phi}(t - t_0)\mathbf{x}_0 + u_s\int_{t_0}^{t}\mathbf{H}\boldsymbol{\Phi}(t - \tau)\mathbf{G}d\tau, \quad t > t_0. \tag{4.14}$$

Let's assume that the input remains constant indefinitely: $u(t) = u_s = const.$, as $t \longrightarrow \infty$. According to Eq. A1.9, the term behind the integral is the impulse–response function (h) of the system with $t \longrightarrow \infty$. Then Eq. 4.14 can only remain bounded if the elements of $\boldsymbol{\Phi}(t - \tau)$

approach zero with $(t - \tau) \longrightarrow \infty$, since otherwise the integral does not stay bounded, since **H** and **G** are constant. This way

$$u_s = u_s \int_{t_0}^{t \to \infty} \mathbf{H}\boldsymbol{\Phi}(t - \tau)\mathbf{G}d\tau = u_s \int_{t_0}^{t \to \infty} h(t - \tau)d\tau \qquad (4.15)$$

can only hold, if the integral in Eq. 4.15 is unity, which means that the system is conservative. This concludes the proof.

A general property of flow-routing models is whether they are *transitive* or not.

Definition 4:   A flow routing model is transitive if the same results is obtained in both cases: (a) the flow is transformed from cross-section L_1 to L_2, and then to L_3; and (b) the flow is transformed in one step from cross-section L_1 to L_3.

Theorem 5:   The $\boldsymbol{\Sigma}_{KMN}$ continuous KMN-cascade is transitive.

Proof: Szöllősi-Nagy (1979) derived the impulse–response of $\boldsymbol{\Sigma}_{KMN}$ by successive convolution, which is based on transitivity.

Note 4.5:   If a system is not conservative, neither is it transitive, because there is a net source or sink in the system.

In this chapter the following conclusions were drawn:

(1) *If a backward difference-scheme is used for spatial differentiation in the partial differential equation of linear kinematic wave, then the so-derived system of ordinary differential equations has a coefficient matrix which is of Toeplitz-band type and its structure is identical to the system matrix of the continuous KMN-cascade.*

(2) *The impulse–response of the continuous, spatially discrete linear kinematic wave is identical to that of the continuous KMN-cascade. Consequently, the two models are equivalent.*

(3) *From (2) follows that the parameters of the two models can be mutually and unambiguously related to each other.*

(4) *In a steady state condition of the continuous KMN-cascade, each storage element contains the same amount of water.*

(5) *The continuous KMN-cascade is transitive.*

EXERCISES

4.1. Demonstrate that the continuous KMN-cascade is indeed transitive for $n = 2$, and then for any n.

4.2. The unit-step ($u_s = 1$ for $t > 0$ and zero otherwise) response function of the continuous KMN cascade is $g(t) = 1 - \sum_{j=0}^{n-1} \frac{(kt)^j}{j!} e^{-kt}$. Since $\delta(t) = \dot{u}_s(t)$, from linearity it follows that $h(t) = \dot{g}(t)$ also. Show that this is true.

State–Space Description of the Discrete Linear Cascade Model (DLCM) and Its Properties: The Pulse-Data System Approach

The practice of operational forecasting requires discrete models because (a) data are generally available at discrete time increments; and (b) forecasting and database models run on digital computers. These two factors fundamentally limit the application of continuous models.

This chapter contains the main results of the study on the deterministic submodel. It specifies conditions necessary for adequate model discretization, namely: discrete coincidence, continuity, and transitivity. Derivation of a discrete state–space model, of which state- and input-transition matrices are in a dual relationship to each other is also included. It demonstrates how different discrete-state representations of the continuous KMN-cascade are related through a linear transformation, and how discrete models are identical to the continuous KMN-cascade in the limit, which means that the discrete models are consistent. It discusses what is meant by the fact that these discrete-state models are discretely coincident with their continuous model counterpart, and, at the same time, illuminates how dynamic changes in the state variable that take place between two adjacent sampling instants are incorporated in the models. It further defines the stability requirements of flow routing as a function of the Courant number. The chapter then focuses on the deterministic prediction of the DLCM state variables, and the determination of the unsteady initial state, required for recursive predictions. Finally, it touches upon the characteristics of the asymptotic behavior of forecasts and upon solving the inverse problem of forecasting, the so-called input detection.

First, however, some results of the not so rare incorrect "trivial" discretization must be mentioned. Models that are discrete by their very nature will not be discussed here (see the works of O'Connor, 1976 and Kontur, 1977 on that subject).

5.1 TRIVIAL DISCRETIZATION OF THE CONTINUOUS KMN-CASCADE AND ITS CONSEQUENCES

Here it is demonstrated why the application of the continuous cascade model in discrete time without modifications to its structure leads to incorrect forecasts.

Let's assume that the continuous input, $u(t)$, and output, $y(t)$, of a continuous linear cascade are *sampled* at *equidistant* time-increments $\Delta t > 0$. Let the so-obtained discrete input and output *time sequences* be u_t and y_t, with discrete time increments $t = 0, \Delta t, 2\Delta t, \ldots$, and so on. (Time will be denoted by a subscript from now on for discrete-time sequences.) The objective is to transform the $\Sigma_{KMN} = (\mathbf{F}, \mathbf{G}, \mathbf{H})$ continuous dynamic model into a discrete-time state–space model

$$\mathbf{x}_{t+\Delta t} = \mathbf{\Phi}(\Delta t)\mathbf{x}_t + \mathbf{\Gamma}(\Delta t)u_t \tag{5.1}$$

$$y_t = \mathbf{H}\mathbf{x}_t \tag{5.2}$$

that meets the criteria of an *adequate discrete representation* as fully as possible. The following definitions are needed to the exact formulation of the problem.

Definition 5:   The $\Sigma_D(\Delta t) = (\mathbf{\Phi}(\Delta t), \mathbf{\Gamma}(\Delta t), \mathbf{H})$ discrete model is *discretely coincident* with the $\Sigma_{KMN} = (\mathbf{F}, \mathbf{G}, \mathbf{H})$ continuous model, if the two model-outputs are identical at discrete time instants of the discrete model and provided the two model inputs are identical at all continuous times.

Definition 6:   A discrete model with equidistant sampling intervals, Δt, of a SISO continuous, linear system is conservative if

$$\sum_{i=1}^{N} h_{i\Delta t} = 1 \tag{5.3}$$

is valid for $N \longrightarrow \infty$, where h, the *unit-pulse response*, is the discrete counterpart of the continuous impulse response function.

Note 5.1:   This definition is analogous to Definition 2 of continuous systems. The unit-pulse function is displayed in Fig. 5.7.

With the help of the above definitions, coupled with Definition 4 (which is model independent, i.e. equally valid for both, continuous and discrete cases), the adequacy of a discrete flow routing model can be defined as:

Definition 7:   The $\Sigma_D(\Delta t) = (\mathbf{\Phi}(\Delta t), \mathbf{\Gamma}(\Delta t), \mathbf{H})$ discrete model defined with equidistant time increments, is a conditionally adequate representation of the $\Sigma_{KMN} = (\mathbf{F}, \mathbf{G}, \mathbf{H})$ continuous model, if it (a) is discretely coincident; (b) keeps its continuity; and (c) is transitive in the $\Delta t \longrightarrow 0$ limit. If (c) is valid for all Δt, then the representation is fully or unconditionally adequate.

Note 5.2:   In the following, $\Delta t = 1$ will be assumed for sake of simplicity. This way the discrete cascade model is written as

$$\mathbf{x}_{t+1} = \mathbf{\Phi}\mathbf{x}_t + \mathbf{\Gamma}u_t \tag{5.4}$$

$$y_t = \mathbf{H}\mathbf{x}_t. \tag{5.5}$$

An exception will be made when the sampling interval has specific importance.

A trivial discretization of the continuous KMN-cascade, Eq. 4.4, is obtained when the system matrices of the discrete model are identical with those of the continuous model. That way the discrete state and output equations become

$$\mathbf{x}_{t+1} = \mathbf{F}\mathbf{x}_t + \mathbf{G}u_t \tag{5.6}$$

$$y_t = \mathbf{H}\mathbf{x}_t. \tag{5.7}$$

Examples for this kind of trivial discretization can be found in Chiu and Isu (1978). This model, $\Sigma'_D = (\mathbf{F}, \mathbf{G}, \mathbf{H})$, however, is not adequate. To prove it, the following is needed

Lemma 2:   If the continuous KMN-cascade, $\Sigma_{KMN} = (\mathbf{F}, \mathbf{G}, \mathbf{H})$, is represented by the $\Sigma'_D = (\mathbf{F}, \mathbf{G}, \mathbf{H})$ discrete model, then the system in its steady state has unequal volumes of water stored in their storage elements

$$x_i = \frac{k^{i-1}}{(1+k)^i}u_s, \quad i = 1, 2, \ldots, n \tag{5.8}$$

where u_s is constant input. When $u_s = 1$, the steady state output is

$$y_s = \frac{k^n}{(1+k)^n} \tag{5.9}$$

which approaches the steady state input ($u_s = 1$) only if $k \longrightarrow \infty$. In that case, however, the total volume of water stored in the cascade approaches zero.

Note 5.3:   For the trivially discretized cascade model to be correct dimensionally, it must be assumed that flow has units of volume, and k is dimensionless.

Proof: A discrete system is in a steady state if

$$\mathbf{x}_s = \mathbf{F}\mathbf{x}_s + \mathbf{G}u_s$$

i.e. the elements of the state variable do not change between two samplings (see Definition 3). Rearrangement of the above equation results in

$$
\begin{aligned}
\mathbf{x}_s &= (\mathbf{I} - \mathbf{F})^{-1}\mathbf{G}u_s \\
&= \mathbf{U}^{-1}\mathbf{G}u_s.
\end{aligned}
$$

The $\mathbf{U} = \mathbf{I} - \mathbf{F}$ matrix can be written as

$$
\mathbf{U} = \begin{bmatrix}
1+k & & & 0 \\
-k & 1+k & & \\
& \ddots & \ddots & \\
0 & & -k & 1+k
\end{bmatrix} = (1+k)\mathbf{I} - k\mathbf{N} = (1+k)(\mathbf{I} - \frac{k}{1+k}\mathbf{N})
$$

so for the inverse it yields

$$\mathbf{U}^{-1} = [(1+k)(\mathbf{I} - \frac{k}{1+k}\mathbf{N})]^{-1} = \frac{1}{1+k}(\mathbf{I} - \frac{k}{1+k}\mathbf{N})^{-1}.$$

The inverse of the $(\mathbf{I} - \frac{k}{1+k}\mathbf{N})$ matrix polynomial can be obtained similarly to the one in Lemma 1:

$$(\mathbf{I} - \frac{k}{1+k}\mathbf{N})^{-1} = \mathbf{I} + \frac{k}{1+k}\mathbf{N} + \frac{k^2}{(1+k)^2}\mathbf{N}^2 + \cdots + \frac{k^{n-1}}{(1+k)^{n-1}}\mathbf{N}^{n-1}.$$

It follows that $\mathbf{U}^{-1}$ is a lower triangular matrix of Toeplitz-type. This way the steady-state system variable is

$$
\mathbf{x}_s = \frac{1}{1+k}
\begin{bmatrix}
1 & 0 & 0 & \cdots & 0 \\
\frac{k}{1+k} & 1 & 0 & \cdots & 0 \\
\frac{k^2}{(1+k)^2} & \ddots & \ddots & \ddots & \vdots \\
\vdots & \ddots & \frac{k}{1+k} & 1 & 0 \\
\frac{k^{n-1}}{(1+k)^{n-1}} & \cdots & \frac{k^2}{(1+k)^2} & \frac{k}{1+k} & 1
\end{bmatrix}
\begin{bmatrix}
1 \\
0 \\
0 \\
\vdots \\
0
\end{bmatrix} u_s
$$

$$
= \frac{1}{1+k}
\begin{bmatrix}
1 \\
\frac{k}{1+k} \\
\frac{k^2}{(1+k)^2} \\
\vdots \\
\frac{k^{n-1}}{(1+k)^{n-1}}
\end{bmatrix} u_s.
$$

The steady-state output becomes

$$y_s = k\frac{1}{1+k}\frac{k^{n-1}}{(1+k)^{n-1}}u_s = \frac{k^n}{(1+k)^n}u_s.$$

Choosing an input of $u_s = 1$ and $n \geq 1$, gives unity only, if $k \longrightarrow \infty$, i.e. the mean storage delay time, $K \longrightarrow 0$, since $K = \frac{1}{k}$. As can be seen, the stored water in the storage elements indeed varies in steady state

$$x_i = \frac{k^{i-1}}{(1+k)^i}u_s, \quad i = 1, 2, \ldots, n$$

and the total water volume, S, in the cascade is

$$S = \sum_{i=1}^{n} \frac{k^{i-1}}{(1+k)^i}u_s = \frac{1}{k}u_s \sum_{i=1}^{n} \frac{k^i}{(1+k)^i}$$

which, with $k \longrightarrow \infty$, becomes

$$\lim_{k \to \infty} S = Ku_s \lim_{k \to \infty} \sum_{i=1}^{n} \frac{k^i}{(1+k)^i} = Ku_s n$$

showing that S indeed approaches zero for a given n. To prove that the above steady-state solution truly represents a steady state,

$$
\mathbf{Fx}_s + \mathbf{Gu}_s =
\begin{bmatrix}
-k & & & 0 \\
k & -k & & \\
 & \ddots & \ddots & \\
0 & & k & -k
\end{bmatrix}
\begin{bmatrix}
\dfrac{1}{1+k} \\[2mm]
\dfrac{k}{(1+k)^2} \\[2mm]
\vdots \\[2mm]
\dfrac{k^{n-1}}{(1+k)^n}
\end{bmatrix}
u_s +
\begin{bmatrix}
1 \\ 0 \\ \vdots \\ 0
\end{bmatrix}
u_s
$$

$$
=
\begin{bmatrix}
\dfrac{-k}{1+k} \\[2mm]
\dfrac{k}{(1+k)^2} \\[2mm]
\vdots \\[2mm]
\dfrac{k^{n-1}}{(1+k)^n}
\end{bmatrix}
u_s +
\begin{bmatrix}
1 \\ 0 \\ \vdots \\ 0
\end{bmatrix}
u_s =
\begin{bmatrix}
\dfrac{1}{1+k} \\[2mm]
\dfrac{k}{(1+k)^2} \\[2mm]
\vdots \\[2mm]
\dfrac{k^{n-1}}{(1+k)^n}
\end{bmatrix}
u_s = \mathbf{x}_s \quad \text{(i)}
$$

can be written.

This concludes the proof.

Theorem 6:   If the continuous KMN-cascade, $\Sigma_{KMN} = (\mathbf{F}, \mathbf{G}, \mathbf{H})$, is represented by the $\Sigma'_D = (\mathbf{F}, \mathbf{G}, \mathbf{H})$ discrete model, then the latter is conservative only if the total water volume stored in the cascade approaches zero.

Proof: Outflow from the discrete cascade at time, τ, is

$$y_\tau \;=\; \mathbf{H}\mathbf{x}_\tau = \mathbf{H}(\mathbf{F}\mathbf{x}_{\tau-1} + \mathbf{G}u_{\tau-1})$$

$$\;=\; \mathbf{H}\mathbf{F}\mathbf{x}_{\tau-1} + \mathbf{H}\mathbf{G}u_{\tau-1}.$$

The second term of the right-hand-side of the equation is zero, because

$$\mathbf{H}\mathbf{G} = [0, 0, \dots, k]\begin{bmatrix} 1 \\ 0 \\ \vdots \\ 0 \end{bmatrix} = 0.$$

In a steady state

$$y_\tau = \mathbf{H}\mathbf{F}\mathbf{x}_s.$$

Applying Eq. (i), gives

$$\mathbf{F}\mathbf{x}_s = \begin{bmatrix} \dfrac{-k}{1+k} \\ \dfrac{k}{(1+k)^2} \\ \vdots \\ \dfrac{k^{n-1}}{(1+k)^n} \end{bmatrix} u_s$$

by which

$$\mathbf{H}\mathbf{F}\mathbf{x}_s = \frac{k^n}{(1+k)^n} u_s = \alpha u_s.$$

As is specified in Corollary 1, the system is conservative if in a steady state

$$y_s = \alpha u_s = u_s$$

which can only happen if $\alpha = 1$. According to Lemma 2, this entails that $k \longrightarrow \infty$, that is, the total volume of water stored in the cascade must approach zero. This concludes the proof.

Corollary 2:   The cascade described by the $\Sigma'_D = (\mathbf{F}, \mathbf{G}, \mathbf{H})$ discrete model is never transitive. This follows from the discrete model being not conservative, due to the presence of artificially introduced net sources or sinks in the discretization scheme (see Note 4.5).

It can be shown through numerical examples that the Σ'_D model does not give identical results to the Σ_{KMN} continuous model at discrete time increments. Consequently, the discrete model is neither discretely coincident.

From the above follows the next:

Theorem 7:   If the continuous KMN-cascade, $\Sigma_{KMN} = (\mathbf{F}, \mathbf{G}, \mathbf{H})$, is represented by the $\Sigma'_D = (\mathbf{F}, \mathbf{G}, \mathbf{H})$ discrete model, then this representation is not adequate.

Note 5.4:   Undoubtedly, the $\Sigma'_D = (\mathbf{F}, \mathbf{G}, \mathbf{H})$ discrete model corresponds to a certain continuous model, but not to the KMN-cascade. Unfortunately, there have been numerous examples of this type of inadequate discretization in the recursive literature in the past. Seeing the unsatisfactory model results, the error has been sought in the estimation algorithms, without realizing that the discrete representation itself was at fault.

5.2 A CONDITIONALLY ADEQUATE DISCRETE MODEL OF THE CONTINUOUS KMN-CASCADE

When instantaneous streamflow measurements (input and output) are only available at discrete time increments, a corresponding discrete state equation must be formulated. Since information on the continuous signal is only available at discrete time increments, some kind of assumption must be made about the behavior of the continuous signal between samples. The two simplest assumptions can be: (a) the signal is constant between subsequent samplings; or (b) the signal changes linearly between discrete sample values. The first approach is called *the pulse-data system approach*, while the second one is called *the linear interpolation (LI) data system approach* (Fig. 5.1). Traditionally, system engineering employed the pulse-data system framework almost exclusively in the past. Consequently, most of the theoretical results involve this approach, which motivated its adoption in water resources applications as well. Derivation of our discrete form of the continuous KMN-cascade below adopts this same framework. However, the results will be reformulated in the next chapter via the application of the LI-data system framework. This latter approach, as will be shown, can be considered as a generalization of the former.

Let's assume that $\mathbf{x}(t)$ is known at time t, and that $\mathbf{u}(t)$ is constant (vector in general) in the (closed from left, open from right) time-interval: $[t, t + \Delta t)$. Then, according to Eq. A1.3,

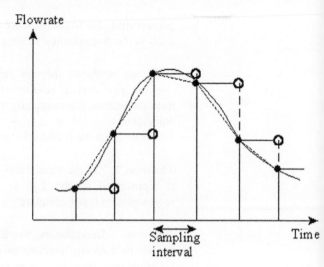

Figure 5.1. Pulse- and LI-data
system representations of a
continuous signal.

$$\mathbf{x}(t + \Delta t) = \mathbf{\Phi}(t + \Delta t, t)\mathbf{x}(t) + [\int_{t}^{t+\Delta t} \mathbf{\Phi}(t + \Delta t, \tau)\mathbf{G}(\tau)d\tau]\mathbf{u}(t)$$

(5.10)

can be written, which can be reformulated with the following definitions:

$$\mathbf{x}_t \triangleq \mathbf{x}(t)$$

$$\mathbf{u}_t \triangleq \mathbf{u}(t)$$

$$\mathbf{\Phi}_t(\Delta t) \triangleq \mathbf{\Phi}(t + \Delta t, t)$$

(5.11)

$$\mathbf{\Gamma}_t(\Delta t) \triangleq \int_{t}^{t+\Delta t} \mathbf{\Phi}(t + \Delta t, \tau)\mathbf{G}(\tau)d\tau$$

(5.12)

as

$$\mathbf{x}_{t+\Delta t} = \mathbf{\Phi}_t(\Delta t)\mathbf{x}_t + \mathbf{\Gamma}_t(\Delta t)\mathbf{u}_t.$$

(5.13)

The discrete output equation, being purely algebraic, remains the same
as in the continuous case

$$\mathbf{y}_t = \mathbf{H}\mathbf{x}_t.$$

(5.14)

5.2.1 *Derivation of the discrete cascade, its continuity, steady state, and transitivity*

The discrete version of the continuous KMN-cascade's state equation
(Eq. 4.4) is

$$\mathbf{x}_{t+\Delta t} = \mathbf{\Phi}(\Delta t)\mathbf{x}_t + \mathbf{\Gamma}(\Delta t)u_t$$

(5.15)

where the state-transition matrix, corresponding to the sampling interval, Δt, becomes

$$\mathbf{\Phi}(\Delta t) \triangleq \mathbf{\Phi}(t + \Delta t, t) = e^{(t+\Delta t - t)\mathbf{F}} = e^{\Delta t \mathbf{F}} \qquad (5.16)$$

(see Eq. A1.6). The input-transition matrix degenerates into a column-vector with a scalar input, $u(t)$,

$$\mathbf{\Gamma}_t(\Delta t) = \int_t^{t+\Delta t} \mathbf{\Phi}(t + \Delta t - \tau)\mathbf{G}(\tau)d\tau \qquad (5.17)$$

and a column-vector $\mathbf{G}$. The discrete model, once again, assumes that the input is constant in the Δt interval: $u(\tau) = const = u_t, \ \tau \ \varepsilon \ [t, t + \Delta t)$.

Note 5.5: Eq. 5.15 provides a discrete description of a continuous process. With those models that are discrete by their very nature, the above derivation of the state and input-transition matrices naturally does not happen because of the lack of a dynamic state change.

The state-transition matrix (Eq. 5.16) that corresponds to Δt, can be obtained from Eq. 4.9 via substituting t with Δt (Szöllősi-Nagy, 1982):

$$\mathbf{\Phi}(\Delta t) = \begin{bmatrix} e^{-\Delta t k} & 0 & 0 & \cdots & 0 \\ \Delta t k e^{-\Delta t k} & e^{-\Delta t k} & 0 & \cdots & 0 \\ \dfrac{(\Delta t k)^2}{2!}e^{-\Delta t k} & \Delta t k e^{-\Delta t k} & e^{-\Delta t k} & 0 & \vdots \\ \vdots & \vdots & \ddots & \ddots & 0 \\ \dfrac{(\Delta t k)^{n-1}}{(n-1)!}e^{-\Delta t k} & \dfrac{(\Delta t k)^{n-2}}{(n-2)!}e^{-\Delta t k} & \cdots & \Delta t k e^{-\Delta t k} & e^{-\Delta t k} \end{bmatrix}$$

$$(5.18)$$

which does not explicitly depend on t, since the model is time-invariant. A useful property of the state-transition matrix is that it always has an inverse (Csáki, 1973); thus $\mathbf{\Phi}(\Delta t)$ is not singular, provided $\Delta t > 0$.

Multiplying the state-transition matrix in Eq. 5.18 by $\mathbf{G}$, from the right, yields the first column of the discrete state-transition matrix at $t + \Delta t - \tau$, which must be integrated over the interval $[t, t + \Delta t)$. The *ith* element of the resulting column-vector is

$$\gamma_{i,1}(\Delta t) = \int_t^{t+\Delta t} \frac{[(t + \Delta t - \tau)k]^{i-1}}{(i-1)!}e^{-(t+\Delta t - \tau)k}d\tau, \quad i = 1, 2, \ldots, n$$

which can be evaluated by the $z = (t + \Delta t - \tau)k$ substitution. The result is a term

$$\gamma_{i,1}(\Delta t) = \frac{1}{k}\frac{1}{(i-1)!}\int_0^{k\Delta t} z^{i-1}e^{-z}dz = \frac{1}{k}\frac{1}{(i-1)!}\Gamma(i,k\Delta t)$$

that contains the incomplete gamma-function, Γ, with parameters: i and $k\Delta t$. It should not be confused with the input-transition matrix, $\boldsymbol{\Gamma}$, which is always denoted by a bold character. Note that for integer values $(i-1)! = \Gamma(i)$, giving

$$\gamma_{i,1}(\Delta t) = \frac{1}{k}\frac{\Gamma(i,k\Delta t)}{\Gamma(i)}, \quad i = 1, 2, \ldots, n. \tag{5.19}$$

In the above expression, the ratio of incomplete and complete gamma functions can be written with the help of Poisson distributions (Rényi, 1968)

$$\frac{\Gamma(i,k\Delta t)}{\Gamma(i)} = 1 - \sum_{j=0}^{i-1} P_j(k\Delta t) \tag{5.20}$$

where

$$P_j(k\Delta t) = \frac{1}{j!}(k\Delta t)^j e^{-k\Delta t} \tag{5.21}$$

is the j-order Poisson distribution with parameter $k\Delta t$. This way the input-transition matrix in Eq. 5.19 has a form (Szöllősi-Nagy, 1982)

$$\boldsymbol{\Gamma}(\Delta t) = \begin{bmatrix} (1 - e^{-\Delta tk})/k \\ [1 - e^{-\Delta tk}(1 + \Delta tk)]/k \\ [1 - e^{-\Delta tk}(1 + \Delta tk + \frac{(\Delta tk)^2}{2})]/k \\ \vdots \\ (1 - e^{-\Delta tk}\sum_{j=0}^{n-1}\frac{(\Delta tk)^j}{j!})/k \end{bmatrix}. \tag{5.22}$$

The state (Eq. 5.18) and input-transition matrix/vector (Eq. 5.22) unambiguously specify the discrete state equation (Eq. 5.15). The discrete output equation remains the same as for the continuous case

$$y_t = \mathbf{H}\mathbf{x}_t. \tag{5.23}$$

As was the case for the continuous model, the discrete model is also unambiguously characterized by the $\boldsymbol{\Sigma}_{DLCM} = (\boldsymbol{\Phi}, \boldsymbol{\Gamma}, \mathbf{H})$ matrix-triplet. Next it is shown that the $\boldsymbol{\Sigma}_{DLCM}$ discrete model is a conditionally adequate representation of the continuous KMN-cascade. For that the following is needed:

Lemma 3:   If the continuous KMN-cascade, $\Sigma_{KMN} = (\mathbf{F}, \mathbf{G}, \mathbf{H})$, is represented by the $\Sigma_{DLCM} = (\boldsymbol{\Phi}, \boldsymbol{\Gamma}, \mathbf{H})$ discrete model, then the steady state of the latter is identical to the steady state of the former.

Proof: The steady-state solution for the continuous case was given by Eq. 4.12

$$
\mathbf{x}_s = \frac{1}{k} \begin{bmatrix} 1 \\ 1 \\ \vdots \\ 1 \end{bmatrix} u_s. \tag{i}
$$

In a steady state the discrete state equation holds for the steady-state solution

$$
\mathbf{x}_s = \boldsymbol{\Phi}\mathbf{x}_s + \boldsymbol{\Gamma}u_s. \tag{ii}
$$

If Eq. (ii) can be shown to hold when Eq. (i) is plugged in for $\mathbf{x}_s$, then the steady-state solution of the continuous model is indeed identical to the steady-state solution of the discrete model. This can be achieved as

$$
\boldsymbol{\Phi}\mathbf{x}_s + \boldsymbol{\Gamma}u_s = e^{-\Delta tk} \begin{bmatrix} 1 & 0 & 0 & \cdots & 0 \\ \Delta tk & 1 & 0 & \cdots & 0 \\ \frac{(\Delta tk)^2}{2!} & \Delta tk & 1 & 0 & \vdots \\ \vdots & \vdots & & \ddots & 0 \\ \frac{(\Delta tk)^{n-1}}{(n-1)!} & \frac{(\Delta tk)^{n-2}}{(n-2)!} & \cdots & \Delta tk & 1 \end{bmatrix} \begin{bmatrix} 1 \\ 1 \\ \vdots \\ 1 \\ 1 \end{bmatrix} \frac{1}{k} u_s + \boldsymbol{\Gamma}u_s
$$

$$
= \frac{1}{k} e^{-\Delta tk} \begin{bmatrix} 1 \\ \Delta tk + 1 \\ \frac{(\Delta tk)^2}{2!} + \Delta tk + 1 \\ \vdots \\ \sum_{j=0}^{n-1} \frac{(\Delta tk)^j}{j!} \end{bmatrix} u_s
$$

$$
+ \begin{bmatrix} (1 - e^{-\Delta tk})/k \\ \left[1 - e^{-\Delta tk}(1 + \Delta tk)\right]/k \\ \left[1 - e^{-\Delta tk}\left(1 + \Delta tk + \frac{(\Delta tk)^2}{2}\right)\right]/k \\ \vdots \\ \left(1 - e^{-\Delta tk}\sum_{j=0}^{n-1} \frac{(\Delta tk)^j}{j!}\right)/k \end{bmatrix} u_s = \frac{1}{k} \begin{bmatrix} 1 \\ 1 \\ \vdots \\ 1 \\ 1 \end{bmatrix} u_s = \mathbf{x}_s
$$

which concludes the proof.

Theorem 8:   If the continuous KMN-cascade, $\Sigma_{KMN} = (\mathbf{F}, \mathbf{G}, \mathbf{H})$, is represented by the $\Sigma_{DLCM} = (\mathbf{\Phi}, \mathbf{\Gamma}, \mathbf{H})$ discrete model, then the latter is conservative.

Proof: The logic is the same as in the proof of Theorem 6. Outflow of the discrete cascade at time, τ, is

$$y_\tau = \mathbf{H}\mathbf{x}_\tau = \mathbf{H}(\mathbf{\Phi}\mathbf{x}_{\tau-1} + \mathbf{\Gamma}u_{\tau-1})$$
$$= \mathbf{H}\mathbf{\Phi}\mathbf{x}_{\tau-1} + \mathbf{H}\mathbf{\Gamma}u_{\tau-1}.$$

In the steady state when $u_\tau = u_s$, $\forall\tau$, Lemma 3 gives

$$\mathbf{x}_s = \frac{1}{k}[1, 1, \ldots, 1]^T u_s.$$

Then,

$$\mathbf{H}\mathbf{\Phi}\mathbf{x}_{\tau-1} =$$

$$[0, 0, \ldots, k]e^{-\Delta tk} \begin{bmatrix} 1 & 0 & 0 & \cdots & 0 \\ \Delta tk & 1 & 0 & \cdots & 0 \\ \dfrac{(\Delta tk)^2}{2!} & \Delta tk & 1 & 0 & \vdots \\ \vdots & \vdots & \ddots & \ddots & 0 \\ \dfrac{(\Delta tk)^{n-1}}{(n-1)!} & \dfrac{(\Delta tk)^{n-2}}{(n-2)!} & \cdots & \Delta tk & 1 \end{bmatrix} \begin{bmatrix} 1 \\ 1 \\ \vdots \\ 1 \\ 1 \end{bmatrix} \frac{1}{k} u_s$$

$$= u_s e^{-\Delta tk} \sum_{j=0}^{n-1} \frac{(\Delta tk)^j}{j!}.$$

Similarly,

$$\mathbf{H}\mathbf{\Gamma}u_{\tau-1} =$$

$$[0, 0, \ldots, k] \begin{bmatrix} (1 - e^{-\Delta tk})/k \\ [1 - e^{-\Delta tk}(1 + \Delta tk)]/k \\ \left[1 - e^{-\Delta tk}\left(1 + \Delta tk + \dfrac{(\Delta tk)^2}{2}\right)\right]/k \\ \vdots \\ \left(1 - e^{-\Delta tk}\sum_{j=0}^{n-1}\dfrac{(\Delta tk)^j}{j!}\right)/k \end{bmatrix} u_s$$

$$= u_s\left(1 - e^{-\Delta tk}\sum_{j=0}^{n-1}\frac{(\Delta tk)^j}{j!}\right).$$

This way

$$y_s = u_s \left[e^{-\Delta t k} \sum_{j=0}^{n-1} \frac{(\Delta t k)^j}{j!} + 1 - e^{-\Delta t k} \sum_{j=0}^{n-1} \frac{(\Delta t k)^j}{j!} \right] = u_s$$

which indeed indicates continuity. This concludes the proof.

Theorem 9:   The continuous KMN-cascade, $\Sigma_{KMN} = (\mathbf{F}, \mathbf{G}, \mathbf{H})$, when represented by the $\Sigma_{DLCM}(\Delta t) = (\mathbf{\Phi}(\Delta t), \mathbf{\Gamma}(\Delta t), \mathbf{H})$ discrete model, keeps its transitivity, provided the sampling interval, $\Delta t \longrightarrow 0$.

Proof: Let the first reach of a stream be bounded by cross-sections L_1 (upstream), and L_2 (downstream), and let's divide the reach into n number of storage elements (Fig. 5.2). Let the second stream reach, consisting of m number of storage elements, be bounded by cross-sections L_2 (upstream), and L_3 (downstream).

For transitivity to hold, it must be proved that the output of the second reach as a response to output of the first reach, is identical to the output of the combined two reaches, taken as one unit. For simplicity, let's consider the case when $n = m = 1$, and the system is relaxed initially, i.e. $\mathbf{x}_0 = \mathbf{0}$. When the two reaches are combined, the discrete output to input first appears at $t = \Delta t$. In the second case, when the two storage elements are considered separate, output of the second storage element is still zero at $t = \Delta t$! The first nonzero output of the second storage element will appear only at $t = 2\Delta t$ to input at $t = \Delta t$, which is the first nonzero output of the first storage element. This immediately proves that the output of the discrete system is generally not the same, depending on whether the system works as one block or as two separate blocks.

Not only the first discrete output value is affected, however. When the system works as one block, the input is transformed between its storage elements according to successive convolution. In our simple example of separate storage elements, the output of the second storage element can be obtained by convolving its unit-pulse response with the output of the first storage element. This is so because the system was assumed to be relaxed. This "theoretical output" of the first storage element will be assumed to be constant during Δt, according to Eq. 5.10, instead of a continuous smooth function of time, as input to the second storage element. (Note that the only difference between the continuous and discrete cascades is

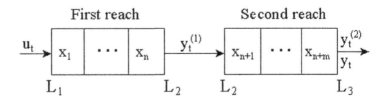

Figure 5.2. Transitivity of the discrete cascade.

in the assumed behavior of the input function. As long as the continuous input function completely matches its assumed behavior during Δt, the discrete model gives identical results to the continuous one at any chosen time.) Consequently, the output of the second storage element must differ from the output of the combined system, because the two inputs to the second storage element are different. Figs. 5.3 and 5.4 demonstrate this concept.

Fig. 5.3 shows the outputs of a system of two storage elements to a constant input of unity with duration Δt (i.e. to a unit-pulse function, see Fig. 5.7) are displayed. At $t = \Delta t$, the output of the separate system is still zero and approaches that of the combined system only as $t \longrightarrow \infty$.

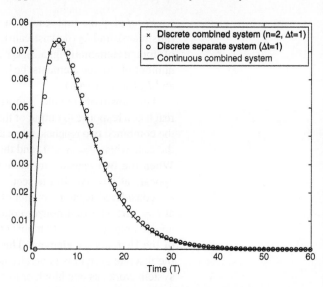

Figure 5.3. Unit-pulse response of a relaxed discrete cascade; (a) as a combined; and (b) as a separate system. $k = 0.2\ [T^{-1}]$.

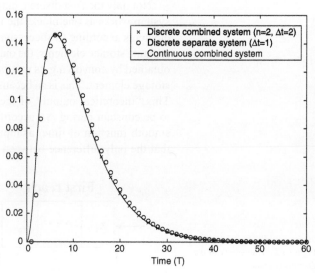

Figure 5.4. Unit-pulse response of a relaxed discrete cascade; (a) as a combined; and (b) as a separate system. $k = 0.2\ [T^{-1}]$.

In Fig. 5.4, the unit-pulse input had a duration of $2\Delta t$ and the combined system had a sampling interval of $2\Delta t$, while the separate system had a sampling interval of Δt. Now, at $t = 2\Delta t$, the output of the separate system is not zero, but it is also different from the combined system's output for the reasons mentioned above.

As $\Delta t \longrightarrow 0$, the difference between the continuous "theoretical output" of a storage element within the cascade and its discrete counterpart tends to zero, due to discrete coincidence. Discrete coincidence directly follows from Eq. 5.15, which is the state trajectory of the continuous KMN-cascade's system equation, taken between two points in time separated by Δt. This means that in the limit, $\Delta t \longrightarrow 0$, the discrete cascade is transitive. This concludes the proof.

The following can now be stated.

Theorem 10: The $\Sigma_{DLCM}(\Delta t) = (\Phi(\Delta t), \Gamma(\Delta t), \mathbf{H})$ discrete model is a conditionally adequate representation of the continuous $\Sigma_{KMN} = (\mathbf{F}, \mathbf{G}, \mathbf{H})$ cascade for stream reaches with no net lateral inflow.

Note 5.6: The discrete model can easily be generalized (Fig. 5.5) for stream reaches having lateral inflow.

The $\mathbf{F}$ state matrix remains the same in the continuous case, and so does the state-transition matrix in the discrete case. If the input of the first storage element of the reach is $u_1(t)$, and the lateral inflows are denoted by $u_j(t)$, $j = 2, \ldots, n$, then the input variable becomes a vector

$$\mathbf{u}(t) = [u_1(t), \ldots, u_n(t)]^T. \tag{5.24}$$

Matrix $\mathbf{G}$ becomes an $n \times n$ identity matrix, and columns of the input-transition matrix, Γ, can be obtained by sliding the vector in Eq. 5.22 along the main diagonal to obtain a lower triangular matrix of Toeplitz-type

$$\Gamma(\Delta t) = \begin{bmatrix} \dfrac{\Gamma(1, \Delta tk)}{k\Gamma(1)} & 0 & \cdots & & 0 \\ \dfrac{\Gamma(2, \Delta tk)}{k\Gamma(2)} & \dfrac{\Gamma(1, \Delta tk)}{k\Gamma(1)} & & 0 & \vdots \\ \vdots & \ddots & & \ddots & 0 \\ \dfrac{\Gamma(n, \Delta tk)}{k\Gamma(n)} & \cdots & & \dfrac{\Gamma(2, \Delta tk)}{k\Gamma(2)} & \dfrac{\Gamma(1, \Delta tk)}{k\Gamma(1)} \end{bmatrix}. \tag{5.25}$$

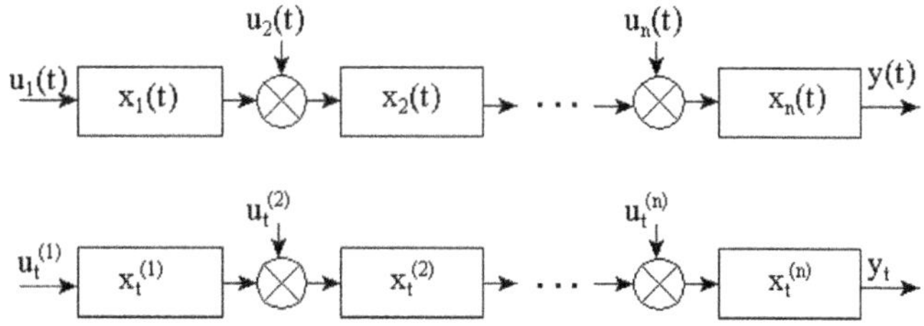

Figure 5.5. Continuous and discrete cascades with lateral inflow.

If there is no lateral inflow at the *ith* position, then the corresponding column in Γ disappears to form an $n \times (n - 1)$ matrix of Toeplitz-type. This keeps repeating with other missing lateral inflows to result in an $n \times 1$ column vector of Eq. 5.22 in the limit of no lateral inflow.

5.2.2 *Relationship between conditionally adequate discrete models with different sampling intervals*

So far the sampling interval, Δt, has been assumed to be set. Let's consider now the case when the discrete model is used with a different sampling interval. A trivial question is if there is any relationship between the two discrete models with different, but constant sampling intervals.

When Δt changes, so do the state-transition matrix (Eq. 5.18) and input-transition (Eq. 5.22) vector. Changing the sampling interval is similar to changing the coordinate system. Provided the discrete model of the continuous KMN-cascade is known for a certain Δt, then the discrete model for any arbitrary Δt^* sampling interval can be derived from it. If the following linear relationship exists between the sampling intervals

$$\Delta t^* = \mu \Delta t, \ \mu \geq 0 \tag{5.26}$$

which is always the case for equidistant samplings, then the system matrices of the new

$$\mathbf{x}_{t+\Delta t^*} = \mathbf{\Phi}(\Delta t^*)\mathbf{x}_t + \mathbf{\Gamma}(\Delta t^*)u_t \tag{5.27}$$

model can be related to the original model through the following

$$\mathbf{\Phi}(\Delta t^*) = \mathbf{T}_\Phi(\mu)\mathbf{\Phi}(\Delta t) \tag{5.28}$$

$$\mathbf{\Gamma}(\Delta t^*) = \mathbf{T}_\Gamma(\mu)\mathbf{\Gamma}(\Delta t) \tag{5.29}$$

linear transformations. Note that μ does not have to be an integer. The lower triangular Toeplitzian $\mathbf{T}_\Phi(\mu)$ transformation matrix can be written as

$$\mathbf{T}_\Phi(\mu) = e^{-\Delta tk(\mu-1)}$$

$$\times \begin{bmatrix} 1 & 0 & 0 & \cdots & 0 \\ \Delta tk(\mu-1) & 1 & 0 & \cdots & 0 \\ \dfrac{[\Delta tk(\mu-1)]^2}{2!} & \Delta tk(\mu-1) & 1 & \cdots & \vdots \\ \vdots & \ddots & \ddots & \ddots & 0 \\ \dfrac{[\Delta tk(\mu-1)]^{n-1}}{(n-1)!} & \dfrac{[\Delta tk(\mu-1)]^{n-2}}{(n-2)!} & \cdots & \Delta tk(\mu-1) & 1 \end{bmatrix} \tag{5.30}$$

while the diagonal $\mathbf{T}_\Gamma(\mu)$ transformation matrix becomes

$$\mathbf{T}_\Gamma(\mu) = < T_{\Gamma 1}, \dots, T_{\Gamma i}, \dots, T_{\Gamma n} > \tag{5.31}$$

with the following diagonal elements

$$T_{\Gamma i} = \frac{\Gamma(i, \Delta t k \mu)}{\Gamma(i, \Delta t k)}. \tag{5.32}$$

From Eq. 5.28 it follows that

$$\mathbf{T}_\Phi(\mu) = \mathbf{\Phi}(\Delta t^*)\mathbf{\Phi}^{-1}(\Delta t). \tag{5.33}$$

The transformation matrix, $\mathbf{T}_\Phi(\mu)$, always exists as it is the state-transition matrix that is invertible for any arbitrary sampling interval, $\Delta t > 0$.

A quick check of the transformation matrix

Example 5.1: Let's show with elementary calculations that the above transformation matrices are correctly specified. First, let's consider a case where input to the cascade becomes zero at time t_0 and remains so afterwards. Let's denote the state of the cascade at time t_0 by $\mathbf{x}_0$. What is its state at $t = t_0 + 2\Delta t$?

According to Eq. 5.15

$$\mathbf{x}_{t_0+\Delta t} = \mathbf{\Phi}(\Delta t)\mathbf{x}_{t_0} \tag{i}$$

$$\mathbf{x}_{t_0+2\Delta t} = \mathbf{\Phi}(\Delta t)\mathbf{x}_{t_0+\Delta t} = \mathbf{\Phi}^2(\Delta t)\mathbf{x}_{t_0}. \tag{ii}$$

At the same time, if $\Delta t^* = 2\Delta t$,

$$\mathbf{x}_{t_0+\Delta t^*} = \mathbf{\Phi}(\Delta t^*)\mathbf{x}_{t_0} \tag{iii}$$

can be written.

If the transformation matrix, $\mathbf{T}_\Phi(\mu)$, is specified correctly above, then from Eqs. (i) and (ii)

$$\mathbf{\Phi}^2(\Delta t) = \mathbf{\Phi}(\Delta t^*) \tag{iv}$$

must hold. Let's see, for example, if $\Phi_{i,i}^2(\Delta t)$ is the same as $\Phi_{i,i}(\Delta t^*)$. From Eqs. 5.28 and 5.30

$$\Phi_{i,i}(\Delta t^*) = e^{-2\Delta t k} \tag{v}$$

follows immediately for $\mu = 2$. Similarly, from Eq. 5.18,

$$\Phi_{i,i}^2(\Delta t) = e^{-2\Delta t k}$$

is obtained.

This indicates that the $\mathbf{T}_\Phi(\mu)$ transformation matrix is given correctly. What about the other transformation matrix, $\mathbf{T}_\Gamma(\mu)$?

Let's assume now, that an initially ($t = 0$) relaxed cascade is fed by a constant inflow of unity with a duration of $2\Delta t$. What is the stored water volume in storage element i at $t = 2\Delta t$?

Water storage in the cascade is again given by Eq. 5.15

$$\mathbf{x}_{\Delta t} = \mathbf{\Gamma}(\Delta t) \tag{vi}$$

$$\mathbf{x}_{2\Delta t} = \mathbf{\Phi}(\Delta t)\mathbf{x}_{\Delta t} + \mathbf{\Gamma}(\Delta t) = \mathbf{\Phi}(\Delta t)\mathbf{\Gamma}(\Delta t) + \mathbf{\Gamma}(\Delta t). \tag{vii}$$

Using the larger sampling interval with $\mu = 2$,

$$\mathbf{x}_{\Delta t^*} = \mathbf{\Gamma}(\Delta t^*) \tag{viii}$$

can be written.

If the transformation matrix, $\mathbf{T}_\Gamma(\mu)$, is given correctly, then Eqs. (vii) and (viii) must be equal. The *ith* element of $\mathbf{\Gamma}(\Delta t^*)$ is given by Eqs. 5.19, 5.29 and 5.31 as

$$\frac{1}{k}\frac{\Gamma(i, 2\Delta tk)}{\Gamma(i, \Delta tk)}\frac{\Gamma(i, \Delta tk)}{\Gamma(i)} = \frac{1}{k}\frac{\Gamma(i, 2\Delta tk)}{\Gamma(i)} \tag{ix}$$

which can be written, using Eqs. 5.20 and 5.21, as

$$\frac{1}{k}\left[1 - \sum_{j=0}^{i-1}\frac{1}{j!}(2\Delta tk)^j e^{-2\Delta tk}\right]. \tag{x}$$

The *ith* element of Eq. (vii) is

$$\frac{1}{k}\left[e^{-\Delta tk}\sum_{j=i-1}^{0}\frac{1}{j!}(\Delta tk)^j\frac{\Gamma(j-i+2, \Delta tk)}{\Gamma(j-i+2)} + \frac{\Gamma(i, \Delta tk)}{\Gamma(i)}\right]. \tag{xi}$$

It is not obvious to see yet that, indeed, Eqs. (x) and (xi) are identical. Let's specify $i = 1$. Then, Eq. (x) becomes

$$\frac{1}{k}(1 - e^{-2\Delta tk}) \tag{xii}$$

while Eq. (xi) simplifies to

$$\frac{1}{k}[e^{-\Delta tk}\frac{\Gamma(1, \Delta tk)}{\Gamma(1)} + \frac{\Gamma(1, \Delta tk)}{\Gamma(1)}] \tag{xiii}$$

which is indeed equal to Eq. (xii), with the help of Eqs. 5.20 and 5.21,

$$\frac{1}{k}[e^{-\Delta tk}(1 - e^{-\Delta tk}) + 1 - e^{-\Delta tk}] \equiv \frac{1}{k}(1 - e^{-2\Delta tk}).$$

This concludes the example.

The above are summarized in the following:

Theorem 11: Any two conditionally adequate representations, belonging to sampling intervals Δt and Δt^*, respectively, of the continuous Σ_{KMN} cascade, are related through a linear transformation

$$\Sigma_{DLCM}(\Delta t) \xrightarrow{\mathbf{T}_\Phi(\mu),\ \mathbf{T}_\Gamma(\mu)} \Sigma_{DLCM}(\Delta t^*) \tag{5.34}$$

where $\mu = \Delta t^*/\Delta t$, and the transformation matrices, $\mathbf{T}_\Phi(\mu)$ and $\mathbf{T}_\Gamma(\mu)$, are defined by Eqs. 5.30 and 5.31.

Note 5.7: When $\mu = 1$, the transformation matrices become the identity matrix. When $\mu \longrightarrow 0$, the discrete model approaches the continuous model, and in the limit they are identical (see Eqs. A1.1, A1.3, and A1.4): $\Sigma_{DLCM}(0) = \Sigma_{KMN}$. This is another proof of consistency of the discretization.

Theorem 12: Any discrete model that is derived from a conditionally adequate discrete model, $\Sigma_{DLCM}(\Delta t)$, via the above transformations, is an equally conditionally adequate model.

Note 5.8: A noteworthy duality can be observed between the state-transition matrix and the input-transition vector. If the order of the cascade, n, is considered a variable, then the first column of the state-transition matrix in Eq. 5.18 contains the impulse responses of those cascades with increasing order (disregarding the multiplier, k). Similarly, the input-transition vector in Eq. 5.22 contains the step responses of those cascades.

The existence of the above linear transformations makes it possible to keep a conditionally adequate discrete model even when the sampling interval is changed, without any need of additional parameter optimization. The forecaster can choose between (a) changing the sampling interval value in the state and input-transition matrices; or (b) leaving the matrices intact, but then they must be multiplied with the corresponding transformation matrices. The fact that the model parameters do not have to be reoptimized may save the user significant computation time.

5.2.3 *Temporal discretization and numerical diffusion*

As was shown in Theorem 3, the linear kinematic wave and continuous KMN-cascade are equivalent. Consequently, discretization results for the latter directly apply for the temporally and spatially discrete linear kinematic wave as well.

Corollary 3: Different discrete representations (i.e. those that correspond to different sampling intervals) of the continuous, spatially discrete linear kinematic wave are related through a linear transformation. The discrete models are not only discretely coincident with the continuous model, but they account for dynamic changes in the modeled process between two sampling instants.

Corollary 4: Temporally and spatially discrete linear kinematic waves belonging to different sampling intervals are related by the same linear transformation, specified in Eq. 5.34, as in the DLCM case.

In connection with spatial discretization, an interesting property must be mentioned, namely: *numerical diffusion*. As the linear kinematic wave is the solution of the pure convection equation, it does not flatten out through time or even change its shape. Rather, the linear kinematic wave simply translates itself from one spatial location to the next (see Eq. 2.14). However, when the linear kinematic wave is discretized either in space (as in the case of the continuous, spatially discrete linear kinematic wave) or directly in space and time (as in the traditional Muskingum model [Ponce, 1980]), using an "off-centered" discretization scheme, it does flatten out (Cunge, 1969). This way, the source of the apparent diffusion is in the numerical scheme itself; that is why this kind of diffusion is referred to as numerical diffusion.

During direct discretization (involving both time and space) of the kinematic wave equation, using "off-centered" differences, the stability of the numerical scheme is conditional. The *Courant-number*,

$$\mathsf{C} = C \frac{\Delta t}{\Delta l} \qquad (5.35)$$

is the parameter that stability depends on. For the numerical scheme to be stable, $\mathsf{C} \leq 2$ condition must be met (Ponce, 1980). Note that this stability criterion is *absent* for the continuous, spatially discrete linear kinematic wave, Eq. 3.7, due to the absence of time differences.

Theorem 13: The discrete linear kinematic wave, $\Sigma_{DLCM}(\Delta t)$, is unconditionally stable numerically.

Proof: As has been shown, the discrete model, $\Sigma_{DLCM}(\Delta t)$, is discretely coincident with the continuous, spatially discrete version, Σ_{KMN}. Solution of the continuous, spatially discrete model does not involve temporal differences. Rather, it is solved via direct integration in time. Discrete coincidence this way assures that stability of the discrete model does not depend on the sampling interval. This concludes the proof.

Corollary 5: Unconditional stability is valid for any sampling interval, $\Delta t^* = \mu \Delta t$, $\mu \geq 0$.

The extent of diffusion in the continuous Σ_{KMN} model is a function of n, the number of storage elements in the stream reach, and K, the mean storage delay time. For a given K, diffusion increases with n, and similarly, for a given n, diffusion increases with K. Fig. 5.6 illustrates this effect for $n = 1, 2, \dots, 6$, where the impulse responses of the continuous KMN-cascade, Eq. 4.8, are plotted.

Note 5.9: The location of the impulse–response function's maximum, the time to peak, t_p, can be calculated by differentiating the impulse–response function with respect to time

$$\dot{h}(t) = \frac{(tk)^{n-1}}{(n-1)!} e^{-tk} \left[\frac{k(n-1)}{t} - k^2 \right] \qquad (i)$$

and solving Eq. (i) for zero, which yields

$$t_p = \frac{n-1}{k}.$$

The peak value is obtained by substituting the t_p value into Eq. 4.8.

Finally, the discrete model is in a form which allows for the application of digital filtering techniques. The discrete model is discretely coincident with its continuous version and is able to account for dynamic changes in the system taking place between samplings. While in the pulse-data

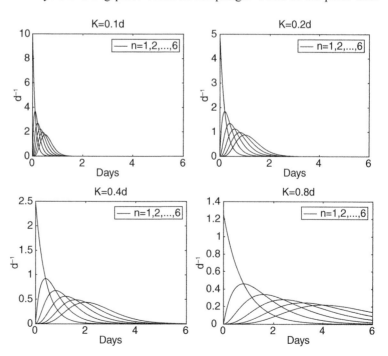

Figure 5.6. Impulse response of the continuous KMN-cascade as a function of n and K.

system framework the input is assumed to be constant between sampling instants, it is not so with the system matrices, for which dynamics over the sampling interval is accounted in the model. The LI-data system approach, discussed later, will also account for dynamic changes in the input variable between discrete samplings.

5.3 DETERMINISTIC PREDICTION OF THE STATE VARIABLES OF THE DISCRETE CASCADE USING A LINEAR TRANSFORMATION

Let $\mathbf{x}_{t+\tau|t}$ denote the *conditional deterministic prediction* of the state variable for time $t + \tau$, with a lead-time of $\tau > 0$, based on information available up to time t. This kind of prediction involves *linear projection* of the *state trajectory*.

At time t, the state variable, $\mathbf{x}_t$, and input, u_t, are available. The one-step forecast, $\Delta t = 1$, derives from the discrete state equation, Eq. 5.15, as

$$\mathbf{x}_{t+1|t} = \mathbf{\Phi}(\Delta t)\mathbf{x}_t + \mathbf{\Gamma}(\Delta t)u_t. \tag{5.36}$$

Note 5.10:   The pulse-data system implicitly assumes that the input, u, at time t will remain constant up to, but not quite reaching, $t + 1$, when it suddenly jumps to its new, future value. This is in accordance with Definition 1 where future estimates, available at time t, are also included among the inputs of the forecasting problem. Inclusion of future estimates of input for forecasting becomes more explicit later, in the LI-data system approach.

The multi-step forecast is formulated in:

Theorem 14:   Deterministic prediction of lead-time $i\Delta t$ $(i > 1)$ of the discrete cascade, $\mathbf{\Sigma}_{DLCM}(\Delta t) = (\mathbf{\Phi}, \mathbf{\Gamma}, \mathbf{H})$, based on information of state, $\mathbf{x}_t$, and input, u_t, variables, is given by the

$$\mathbf{x}_{t+i\Delta t|t} = \mathbf{\Phi}[(i - 1)\Delta t]\mathbf{x}_{t+\Delta t|t} \tag{5.37}$$

linear transformation, where

$$\mathbf{x}_{t+\Delta t|t} = \mathbf{\Phi}(\Delta t)\mathbf{x}_t + \mathbf{\Gamma}(\Delta t)u_t$$

and $u_{t+i\Delta t} = 0$, $i > 0$ is assumed.

Proof: By definition, the state-transition matrix is

$$\mathbf{\Phi}(\Delta t) = \mathbf{\Phi}(t + \Delta t, t)$$

by which

$$\mathbf{x}_{t+\Delta t|t} = \mathbf{\Phi}(t + \Delta t, t)\mathbf{x}_t + \mathbf{\Gamma}(\Delta t)u_t. \tag{i}$$

The state at $t + 2\Delta t$, using the state at $t + \Delta t$, can be expressed as a linear transformation, provided $\mathbf{x}_{t+\Delta t|t}$ has been estimated and $u_{t+\Delta t|t} = 0$:

$$\mathbf{x}_{t+2\Delta t|t} = \mathbf{\Phi}(t + 2\Delta t, t + \Delta t)\mathbf{x}_{t+\Delta t|t}. \tag{ii}$$

For $t + \Delta t$, Eq. (i) gives a deterministic forecast, which upon substitution into Eq. (ii), results in a multi-step forecast from time t. Similarly, the state at $t + 3\Delta t$, can be predicted from $t + 2\Delta t$, as

$$\mathbf{x}_{t+3\Delta t|t} = \mathbf{\Phi}(t + 3\Delta t, t + 2\Delta t)\mathbf{x}_{t+2\Delta t|t}$$

which after insertion of Eq. (ii) yields

$$\mathbf{x}_{t+3\Delta t|t} = \mathbf{\Phi}(t + 3\Delta t, t + 2\Delta t)\mathbf{\Phi}(t + 2\Delta t, t + \Delta t)\mathbf{x}_{t+\Delta t|t}$$
$$= \mathbf{\Phi}(t + 3\Delta t, t + \Delta t)\mathbf{x}_{t+\Delta t|t}$$

where the following chain property of the state-transition matrix was exploited:

$$\mathbf{\Phi}(t_3, t_1) = \mathbf{\Phi}(t_3, t_2)\mathbf{\Phi}(t_2, t_1). \tag{5.38}$$

In general, the following is obtained

$$\mathbf{x}_{t+i\Delta t|t} = \mathbf{\Phi}(t + i\Delta t, t + \Delta t)\mathbf{x}_{t+\Delta t|t} \tag{iii}$$

where

$$\mathbf{\Phi}(t + i\Delta t, t + \Delta t) = \prod_{j=1}^{i-1} \mathbf{\Phi}(t + (j + 1)\Delta t, t + j\Delta t). \tag{iv}$$

Being the cascade model time-invariant (see Eq. 5.16) and the discretization equidistant,

$$\mathbf{\Phi}(t + (j + 1)\Delta t, t + j\Delta t) = \mathbf{\Phi}(\Delta t) \tag{v}$$

and

$$\mathbf{\Phi}(t + i\Delta t, t + \Delta t) = \mathbf{\Phi}[(i - 1)\Delta t)] \tag{vi}$$

which concludes the proof.

Note 5.11:   Eq. (iv) shows that for a time-invariant model with equidistant sampling-interval

$$[\Phi(\Delta t)]^{i-1} = \Phi[(i-1)\Delta t].$$ (5.39)

This means that in recursive predictions the potentially time-consuming matrix power function can be replaced by a simple change of the multiplier of Δt in the matrix elements. This property follows from the identity

$$[\Phi(\Delta t)]^{i-1} = [e^{\Delta t \mathbf{F}}]^{i-1} = e^{(i-1)\Delta t \mathbf{F}}.$$ (5.40)

Note 5.12:   The forecasting equation (Eq. 5.37) is valid only if it is assumed that $u_{t+i\Delta t} = 0$ for $i > 0$. The state prediction formula in Eq. 5.37 is really the homogeneous solution of the discrete state equation, Eq. 5.15, for $i > 1$. The accuracy of the deterministic forecast can only be increased if information is available on the future expected value of the input, which can be a forecast for the stream reach upstream.

5.4 CALCULATION OF SYSTEM CHARACTERISTICS

System-characteristic matrices are those matrices that relate the input of a system to its output, and consequently, system output can be specified to any arbitrary input with their help. The matrices become single-valued functions of time in the continuous case and sequences for discrete systems when the system is SISO, i.e. the input and output are both scalars. These characteristic matrices, or functions, if we stay with the SISO system framework, are in fact system outputs to well-defined special inputs, and as such they implicitly contain all the properties characteristic of the system. In time-domain analysis, the two characteristic functions are the *impulse response*, which is the system output to input in the form of a Dirac-delta function, and *the unit-step-response functions*, the system response to an input in the form of a unit-step function. In section 4.2, it was mentioned that the impulse response of the continuous KMN-cascade can be calculated from the matrix-triplet $(\mathbf{F}, \mathbf{G}, \mathbf{H})$ as Eq. 4.8. Due to the integral/differential relationship between the Dirac-delta and unit-step functions, the impulse–response function can be obtained by differentiating the unit-step-response function. This, however, is not trivial in the discrete case, when the continuous characteristic functions are interpreted only in discrete time instants, and so they cannot be differentiated in the traditional sense. However, the discrete characteristics can be calculated straightforwardly from the solution of the discrete state equation.

In Theorem 14, it was shown how the discrete states of the homogeneous system can be simply calculated by recursive substitution. The same is true for the inhomogeneous case. Assuming that the initial state, $\mathbf{x}_0$, and the input sequence, $u_{i\Delta t}$, are known for $i = 0, 1, \ldots, n-1$, then the corresponding states can be obtained as

$$\mathbf{x}_{\Delta t} = \Phi(\Delta t)\mathbf{x}_0 + \Gamma(\Delta t)u_0$$

$$\mathbf{x}_{2\Delta t} = \Phi(\Delta t)\mathbf{x}_{\Delta t} + \Gamma(\Delta t)u_{\Delta t}$$

$$= \Phi^2(\Delta t)\mathbf{x}_0 + \Phi(\Delta t)\Gamma(\Delta t)u_0 + \Gamma(\Delta t)u_{\Delta t}$$

$$\vdots$$

$$\mathbf{x}_{n\Delta t} = \Phi^n(\Delta t)\mathbf{x}_0 + \sum_{i=0}^{n-1} \Phi^{n-i-1}(\Delta t)\Gamma(\Delta t)u_{i\Delta t}. \tag{5.41}$$

From this it can be seen that the solution consists of two parts: the first term describes the effect of the initial condition, while the second term specifies the effect of the inputs to the development of the state. (Compare it with the continuous case, Eq. A1.3.)

As has been mentioned earlier, the most important advantage of the application of DLCM lies in its recursivity, which may distinguish it from other hydrological forecasting models. However, the discrete system-characteristic functions, at least in the pulse-data framework, become cardinal in the computation of the unsteady initial condition, which is not at all a trivial problem. Therefore, the discrete counterparts of the impulse and unit-step-response functions of the KMN-cascade will be discussed below.

5.4.1 *Unit-pulse response of the discrete cascade*

In discrete time, the Dirac-delta function becomes the *unit-pulse sequence*, defined as

$$\delta_{i\Delta t} = 1, \ i = 0 \tag{5.42}$$

$$= 0, \ i = 1, 2, \dots .$$

Fig. 5.7 illustrates the resultant *unit-pulse function*, $u_p(t)$, within the pulse-data system framework.

The unit-pulse response of a discrete, linear, time-invariant system can be obtained similar to the continuous case, but not in an identical way

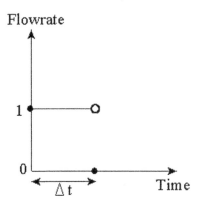

Figure 5.7. Interpretation of the unit-pulse function, $u_p(t)$, in the pulse-data system framework.

(see Eq. A1.11). Let the system be relaxed initially, $\mathbf{x}_0 = \mathbf{0}$. The output, according to Eqs. 5.23 and 5.41, is

$$y_{i\Delta t} = \mathbf{H}\mathbf{x}_{i\Delta t} = \sum_{j=0}^{i-1} \mathbf{H}\mathbf{\Phi}^{i-j-1}(\Delta t)\mathbf{\Gamma}(\Delta t)u_{j\Delta t} \tag{5.43}$$

which is indeed a *discrete convolution*, where the

$$\mathbf{H}\mathbf{\Phi}^{i-1}(\Delta t)\mathbf{\Gamma}(\Delta t), \quad i \geq 1 \tag{5.44}$$

triple-product is the unit-pulse response at discrete time instants, $i\Delta t$, $i \geq 1$. That it is so can be seen by the convolution of the unit-pulse input and the above expression

$$y_{i\Delta t} = \sum_{j=0}^{i-1} \mathbf{H}\mathbf{\Phi}^{i-j-1}(\Delta t)\mathbf{\Gamma}(\Delta t)\delta_{j\Delta t} = \mathbf{H}\mathbf{\Phi}^{i-1}(\Delta t)\mathbf{\Gamma}(\Delta t) = h_{i\Delta t}, \ i \geq 1$$

$$\tag{5.45}$$

which indeed gives back Eq. 5.44. According to Eq. 5.39 the unit-pulse response can be written as

$$h_{i\Delta t} \equiv h_i(\Delta t) = \mathbf{H}\mathbf{\Phi}[(i-1)\Delta t]\mathbf{\Gamma}(\Delta t), \ i \geq 1. \tag{5.46}$$

Note 5.13: The discrete unit-pulse response, $h_i(\Delta t)$ ($\equiv h_{i\Delta t}$), is not specified at $t = 0$, due to the discrete nature of the model. This means that the effect of any disturbance of the system (e.g. at time $t = 0$) can show up in the output only Δt time later, i.e.

$$y_{\Delta t} = \mathbf{H}\mathbf{x}_{\Delta t} = \mathbf{H}[\mathbf{\Phi}(\Delta t)\mathbf{x}_0 + \mathbf{\Gamma}(\Delta t)u_0]$$

From this it follows that the discrete model is a delayed-response system, in opposition to the continuous model.

Note 5.14: The $h_i(\Delta t)$ unit-pulse response unambiguously specifies a discrete system within the pulse-data system. From this it follows that the $[\mathbf{\Phi}(\Delta t), \mathbf{\Gamma}(\Delta t), \mathbf{H}]$ matrix-triplet unambiguously characterizes a discrete linear, time-invariant, dynamic system.

Theorem 15: The unit-pulse response of the $\mathbf{\Sigma}_{DLCM}(\Delta t) = [\mathbf{\Phi}(\Delta t), \mathbf{\Gamma}(\Delta t), \mathbf{H}]$ n-order discrete cascade is given by

$$h_{i\Delta t} = e^{-(i-1)\Delta tk}\left[\sum_{j=1}^{n} \frac{[(i-1)\Delta tk]^{n-j}}{(n-j)!}\left(1 - e^{-\Delta tk}\sum_{m=0}^{j-1}\frac{(\Delta tk)^m}{m!}\right)\right]$$

$$n \geq 1, \ k > 0, \ \Delta t > 0, \ i = 1, 2, \ldots . \tag{5.47}$$

Proof: As $\mathbf{H} = [0, 0, \ldots, k]$, it picks out the $\boldsymbol{\Phi}(\cdot)\boldsymbol{\Gamma}$ product's last element, which is the

$$e^{-(i-1)\Delta tk} \left[\frac{[(i-1)\Delta tk]^{n-1}}{(n-1)!}, \ldots, \frac{[(i-1)\Delta tk]^{n-j}}{(n-j)!}, \ldots, 1 \right]$$

$$\times \begin{bmatrix} (1 - e^{-\Delta tk})/k \\ [1 - e^{-\Delta tk}(1 + \Delta tk)]/k \\ \left[1 - e^{-\Delta tk}\left(1 + \Delta tk + \frac{(\Delta tk)^2}{2}\right)\right]/k \\ \vdots \\ \left(1 - e^{-\Delta tk}\sum_{j=0}^{n-1}\frac{(\Delta tk)^j}{j!}\right)/k \end{bmatrix}$$

scalar product. Multiplying this by k gives Eq. 5.47, which concludes the proof.

Note 5.15:   The unit-pulse response satisfies the following equality

$$\lim_{N \to \infty} \sum_{i=1}^{N} h_{i\Delta t} = 1, \quad \forall(n \geq 1, \, k > 0, \, \Delta t > 0) \tag{5.48}$$

for equidistant sampling. This follows from the $\boldsymbol{\Sigma}_{DLCM}(\Delta t)$ discrete cascade's property of being conservative (see Theorem 8). As the unit-pulse response is the outflow of an initially relaxed system to inflow in the shape of a unit-pulse function, as time approaches infinity, the total inflow must equal the total outflow if the system is conservative. This means that

$$\int_{t=0}^{\infty} u(t)dt = \int_{t=0}^{\infty} u_p(t)dt = \Delta t \sum_{i=0}^{N \to \infty} \delta_{i\Delta t} = \Delta t$$

must equal

$$\int_{t=0}^{\infty} y(t)dt = \int_{t=0}^{\infty} h_p(t)dt = \Delta t \sum_{i=1}^{N \to \infty} h_{i\Delta t} \tag{i}$$

which can only be true if the right-hand-side of Eq. (i) sums to unity. Recall that in the pulse-data system any sampled function is assumed to have a constant value, equal to the last sampling value, during the sampling interval. Here $h_p(t)$ denotes the continuous function obtained from discrete values, $h_{i\Delta t}$, within the pulse-data framework, similar to the unit-pulse function interpretation.

Another property of the unit-pulse response is that

$$\lim_{i \to \infty} h_i(\Delta t) = 0$$

and

$$h_i(\Delta t) \geq 0, \quad \forall i \in (1, 2, \ldots). \tag{5.49}$$

The above three properties correspond to

$$\int_{-\infty}^{\infty} h(\tau) d\tau = 1$$

$$\lim_{\tau \to \infty} h(\tau) = 0$$

$$h(\tau) \geq 0$$

in the continuous case (Diskin and Boneh, 1972).

Note 5.16:   From Eq. 5.47, the unit-pulse-response value at $t = \Delta t$ is

$$h_1(\Delta t) = 1 - e^{-\Delta t k} \sum_{m=0}^{n-1} \frac{(\Delta t k)^m}{m!} \tag{5.50}$$

which is the last element of the input-transition matrix times k. The same must be obtained by Eq. 5.41 with $\mathbf{x}_0 = 0$ and $u_0 = 1$:

$$\mathbf{x}_{\Delta t} = \mathbf{\Phi} \mathbf{x}_0 + \mathbf{\Gamma} 1 = \mathbf{\Gamma}$$

and

$$y_{\Delta t} = \mathbf{H} \mathbf{x}_{\Delta t} = \mathbf{H} \mathbf{\Gamma} = k \gamma_n$$

as $u_{i\Delta t} = 0$, $i = 1, 2, \ldots$.

Discrete coincidence demonstration

Example 5.2:   Let $\Delta t = 1d$, $n = 1$, and $K = 3d$. From Eq. 5.47 the unit-pulse response is

$$h_i(1) = e^{-(i-1)k}(1 - e^{-k}).$$

Fig. 5.8 displays the corresponding unit-pulse-response sequence together with the continuous convolution result using Eqs. 4.8 and A1.10. Discrete coincidence is obvious.

The same unit-pulse response can, of course, be obtained from Eq. 5.41 with

$$\mathbf{\Phi} = e^{-k}; \quad \mathbf{\Gamma} = \frac{1}{k}(1 - e^{-k}); \quad u_{i\Delta t} = \delta_{i\Delta t}; \quad H = k$$

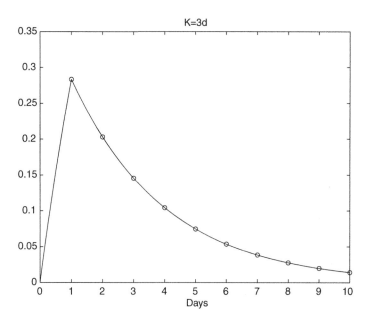

Figure 5.8. Unit-pulse responses of the discrete (circles) and continuous cascade models.

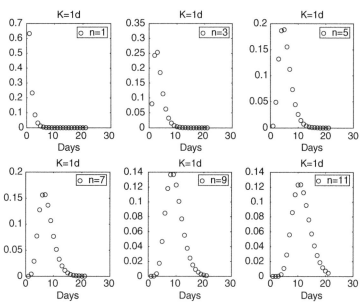

Figure 5.9. Unit-pulse response of the discrete linear cascade with increasing number of the storage elements. $\Delta t = 1$ day.

as

$$h_i(1) = H\Phi^{(t-1)}\Gamma = ke^{-(i-1)k}\frac{1}{k}(1 - e^{-k}) = e^{-(i-1)k}(1 - e^{-k}).$$

Fig. 5.9 illustrates the effect of the increasing number of storage elements, n, on the unit-pulse response of the discrete model. These results (see Note 5.9) are similar to Fig. 5.6 in the sense that the unit-pulse-response ordinates decrease, while the time to peak increases with increasing value of the mean delay time of the reach, nK.

Note 5.17:   The time to peak can be easily calculated for the continuous unit-pulse-response function. The unit-pulse function can be written as

$$u_p(t) = 1(t) - 1(t - \Delta t)$$

where $1(t)$ is the continuous unit-step function

$$1(t) \quad = \quad 0, \quad t < 0$$
$$= \quad 1, \quad t \geq 0.$$

Due to linearity, the unit-pulse-response function, $h_p(t)$, can also be obtained as the difference in the unit-step-response functions, $g(t)$,

$$h_p(t) = g(t) - g(t - \Delta t)$$

with $g(t)$ specified in Table 5.1. This way $h_p(t)$ becomes

$$h_p(t) = e^{-kt} \sum_{j=0}^{n-1} \left[\frac{[(t - \Delta t)k]^j}{j!} e^{\Delta tk} - \frac{(tk)^j}{j!} \right], \quad t \geq \Delta t$$

which upon differentiation with respect to time and solving for zero yields for the time to peak

$$t_p \quad = \quad \frac{\Delta t e^{\frac{k}{n-1}}}{e^{\frac{k}{n-1}} - 1}, \quad n > 1$$
$$= \quad \Delta t, \quad n = 1.$$

From this, the time to peak for the discrete unit-pulse responses in Fig. 5.9 can be obtained as

$$t_p(\Delta t) = h_p^{-1}(\max[h_p(int\{t_p - 0.5\Delta t\}), h_p(int\{t_p + 0.5\Delta t\})])$$

which simply states that the discrete peak to time value results at the discrete time instant where the continuous unit-pulse-response function has a maximum among the two discrete time-instants that enclose t_p. Calculation of the discrete time to peak can be done this way because the discrete unit-pulse response is discretely coincident with the continuous unit-pulse function at discrete time-increments.

5.4.2 *Unit-step response of the discrete cascade*

The unit-step sequence is defined as

$$1_{i\Delta t} = 0, \quad i < 0 \tag{5.51}$$
$$= 1, \quad i \geq 0.$$

The unit-step response is the initially relaxed system's output to a unit-step input, i.e. (see Eq. 5.41):

$$\mathbf{x}_1 = \mathbf{\Phi}\mathbf{x}_0 + \mathbf{\Gamma}\mathbf{1} = \mathbf{\Gamma}$$
$$\mathbf{x}_2 = \mathbf{\Phi}\mathbf{x}_1 + \mathbf{\Gamma}\mathbf{1} = \mathbf{\Phi}\mathbf{\Gamma} + \mathbf{\Gamma}$$
$$\vdots$$
$$\mathbf{x}_N = \left(\sum_{i=0}^{N-1} \mathbf{\Phi}^i\right)\mathbf{\Gamma} \tag{5.52}$$

with a notation involving $\Delta t = 1$. The system output using a sampling interval of Δt is

$$y_{N\Delta t} = \mathbf{H}\left[\sum_{i=0}^{N-1} \mathbf{\Phi}^i(\Delta t)\right]\mathbf{\Gamma}(\Delta t) \tag{5.53}$$

which, with respect to Eq. 5.39, can be written as

$$g_{N\Delta t} \equiv g_N(\Delta t) = \mathbf{H}\left[\sum_{i=0}^{N-1} \mathbf{\Phi}(i\Delta t)\right]\mathbf{\Gamma}(\Delta t) \tag{5.54}$$

which is the unit-step response of the discrete linear cascade. Note that $\mathbf{\Phi}(0) = \mathbf{I}$, the identity matrix (see Eq. A1.4).

Theorem 16:   The unit-step response of $\mathbf{\Sigma}_{DLCM}(\Delta t) = [\mathbf{\Phi}(\Delta t), \mathbf{\Gamma}(\Delta t), \mathbf{H}]$ for a cascade of $n \geq 1$ order, $k > 0$, and $\Delta t > 0$, is given by

$$g_{i\Delta t} = \sum_{j=1}^{n} \left[\frac{(\Delta tk)^{n-j}}{(n-j)!} \left(\sum_{l=1}^{i}(l-1)^{n-j}e^{-(l-1)\Delta tk}\right) \right.$$
$$\left. \times \left(1 - e^{-\Delta tk}\sum_{m=0}^{j-1}\frac{(\Delta tk)^m}{m!}\right) \right] \quad i = 1, 2, \ldots. \tag{5.55}$$

Proof: It is enough to consider the last row of the matrix sum, $\sum_{i=0}^{N-1} \mathbf{\Phi}(i\Delta t)$, similar to the proof of Theorem 15. Starting with $i = 0$, the last rows in the sum (see Eq. 5.18) are:

$$[0, 0, \ldots, 1]$$

$$\left[\frac{(\Delta tk)^{n-1}}{(n-1)!} e^{-\Delta tk}, \frac{(\Delta tk)^{n-2}}{(n-2)!} e^{-\Delta tk}, \ldots, e^{-\Delta tk} \right]$$

$$\vdots$$

$$\left[\frac{[(N-1)\Delta tk]^{n-1}}{(n-1)!} e^{-(N-1)\Delta tk}, \frac{[(N-1)\Delta tk]^{n-2}}{(n-2)!} e^{-(N-1)\Delta tk}, \ldots, e^{-(N-1)\Delta tk} \right]$$

the sum of which gives the last row of the matrix sum. This is given by

$$\left[\sum_{i=0}^{N-1} \Phi(i\Delta t) \right]_{n,.} = \left[\frac{(\Delta tk)^{n-1}}{(n-1)!} \sum_{i=1}^{N}(i-1)^{n-1} e^{-(i-1)\Delta tk}, \ldots, \right.$$

$$\left. \frac{(\Delta tk)^{n-j}}{(n-j)!} \sum_{i=1}^{N}(i-1)^{n-j} e^{-(i-1)\Delta tk}, \ldots, \sum_{i=1}^{N} e^{-(i-1)\Delta tk} \right]. \tag{i}$$

The right-hand-side of Eq. (i) is multiplied by $\Gamma(\Delta t)$, which, when further multiplied by k, yields Eq. 5.55. This concludes the proof.

Note 5.18:   The discrete unit-step response is zero at $t = 0$, its value at $t = \Delta t$ is given by Eq. 5.55 as

$$g_1(\Delta t) = 1 - e^{-\Delta tk} \sum_{i=0}^{n-1} \frac{(\Delta tk)^i}{i!} \tag{5.56}$$

which is identical to the unit-pulse response value at $t = \Delta t$ (see Eq. 5.50), since up until Δt the unit-step and unit-pulse functions are identical, with the exception of the sampling-instant value at $t = \Delta t$, when the latter becomes zero instantly.

Note 5.19:   The unit-step response of the discrete cascade model with given parameters can be calculated by Eq. 5.55. However, the unit-step response can be easily calculated, provided the unit-pulse response is known. This is because from Eq. 5.46

$$\sum_{i=1}^{N} h_i(\Delta t) = \sum_{i=1}^{N} \mathbf{H}\Phi^{i-1}(\Delta t)\Gamma(\Delta t) = \sum_{i=0}^{N-1} \mathbf{H}\Phi^i(\Delta t)\Gamma(\Delta t)$$

$$= \mathbf{H} \left[\sum_{i=0}^{N-1} \Phi^i(\Delta t) \right] \Gamma(\Delta t)$$

can be written which is Eq. 5.54. This way

$$g_{i\Delta t} \equiv g_i(\Delta t) = \sum_{j=1}^{i} h_{j\Delta t} \qquad (5.57)$$

that is the discrete unit-step response at time t is the sum of the discrete unit-pulse responses up until time t. This relationship is the discrete version of the

$$g(t) = \int_{-\infty}^{t} h(\tau)d\tau \qquad (5.58)$$

integral relationship of continuous systems (Fodor, 1967). This is not surprising given that the Dirac-delta function is the derivative of the continuous unit-step function.

Note 5.20:   From Eqs. 5.48, 5.49 and 5.57 it follows that

$$\lim_{i \to \infty} g_i(\Delta t) \quad = \quad 1, \quad \forall(n \geq 1, k > 0, \Delta t > 0) \qquad (5.59)$$

$$g_i(\Delta t) \quad \geq \quad 0, \quad \forall i \in (1, 2, \dots). \qquad (5.60)$$

The system characteristics of the continuous KMN and discrete DLCM cascades are summarized in Table 5.1.

Table 5.1. System characteristics of the continuous KMN-cascade and DLCM.

KMN

$$h(t) = k\frac{(tk)^{n-1}}{(n-1)!}e^{-tk}$$

$$g(t) = 1 - \sum_{j=0}^{n-1}\frac{(tk)^j}{j!}e^{-tk}$$

$$\int_{-\infty}^{\infty} h(\tau)d\tau = 1, \quad \lim_{\tau \to \infty} h(\tau) = 0, \quad h(\tau) \geq 0, \quad g(t) = \int_{-\infty}^{t} h(\tau)d\tau, \quad g(t) \geq 0$$

DLCM

$$h_{i\Delta t} = e^{-(i-1)\Delta tk}\left[\sum_{j=1}^{n}\frac{[(i-1)\Delta tk]^{n-j}}{(n-j)!}\left(1 - e^{-\Delta tk}\sum_{m=0}^{j-1}\frac{(\Delta tk)^m}{m!}\right)\right]$$

$$g_{i\Delta t} = \sum_{j=1}^{n}\left[\frac{(\Delta tk)^{n-j}}{(n-j)!}\left(\sum_{l=1}^{i}(l-1)^{n-j}e^{-(l-1)\Delta tk}\right)\left(1 - e^{-\Delta tk}\sum_{m=0}^{j-1}\frac{(\Delta tk)^m}{m!}\right)\right]$$

$$\lim_{N \to \infty}\sum_{i=1}^{N} h_{i\Delta t} = 1, \quad \lim_{i \to \infty} h_i(\Delta t) = 0, \quad h_i(\Delta t) \geq 0, \quad g_{i\Delta t} = \sum_{j=1}^{i} h_{j\Delta t}, \quad g_i(\Delta t) \geq 0$$

5.5 CALCULATION OF INITIAL CONDITIONS FOR THE DISCRETE CASCADE

Recursive forecasting, Eqs. 5.15 and 5.41, requires the initial condition, x_0, to be specified. Here it is shown how easily the state–space approach can be used to calculate initial conditions. This is in stark contrast to the input–output convolution model practice, where this has always posed a difficult problem (Kucsment, 1967) and was solved using approximations. To avoid oscillations in the impulse–response function (also called *instantaneous unit hydrograph*), Kucsment (1967) suggested the application of the hard-to-apply regularization technique of Tyhonov. Okunishi (1973) showed that the regularization technique, as a payoff for its difficulty, gives more accurate results than estimation of the impulse–response values using least-squares. In order to circumvent the numerical problems encountered during determination of the initial condition, Hovsepian and Nazarian (1969) used an analog computer. Today, this may seem an archaic approach.

During steady state, according to Lemma 3,

$$x_0 = \frac{1}{k}[1, 1, \ldots, 1]^T u_s \tag{5.61}$$

where u_s is the steady state input. In this case all components of the initial condition vector are equal.

In an unsteady flow condition, components of x_0 have different values, i.e. the storage elements contain different volumes of water. Below it is shown that the n-dimensional vector, x_0, can be specified unambiguously from n number of input–output value pairs. It, however, requires the following:

Theorem 17: The discrete cascade, $\Sigma_{DLCM}(\Delta t)$, is observable, if $n \geq 1$, $k > 0$, and $\Delta t > 0$.

Proof: A time-invariant, discrete, linear dynamic system is observable (Kalman, 1961) if the *observability matrix*

$$\Theta_n = [H\Phi, H\Phi^2, \ldots, H\Phi^n]^T$$

has rank n (for a slightly different definition of the observability matrix, see Eq. A1.18). This means that the rows/columns of Θ_n are linearly independent. The matrix series, $H\Phi^i$ ($i = 1, \ldots, n$), yields the Φ^i matrix-exponential's last row (times k) due to the structure of the row vector H. According to Eq. 5.39, $\Phi^i(\Delta t) = \Phi(i\Delta t)$, consequently, the rows of the $n \times n$ observability matrix

$$\Theta_n = \begin{bmatrix} k\dfrac{(\Delta tk)^{n-1}}{(n-1)!}e^{-\Delta tk} & k\dfrac{(\Delta tk)^{n-2}}{(n-2)!}e^{-\Delta tk} & \cdots & ke^{-\Delta tk} \\ k\dfrac{(2\Delta tk)^{n-1}}{(n-1)!}e^{-2\Delta tk} & k\dfrac{(2\Delta tk)^{n-2}}{(n-2)!}e^{-2\Delta tk} & \cdots & ke^{-2\Delta tk} \\ \vdots & \vdots & k\dfrac{(i\Delta tk)^{n-j}}{(n-j)!}e^{-i\Delta tk} & \vdots \\ k\dfrac{(n\Delta tk)^{n-1}}{(n-1)!}e^{-n\Delta tk} & k\dfrac{(n\Delta tk)^{n-2}}{(n-2)!}e^{-n\Delta tk} & \cdots & ke^{-n\Delta tk} \end{bmatrix}$$

$$(5.62)$$

are all linearly independent from each other, unless $k = 0$. This latter parameter, however, is never zero in a physical sense; thus the discrete cascade is observable (Szöllősi-Nagy, 1987). If $\Delta t = 0$, then the rows become identical; thus the discrete cascade is then not observable. This concludes the proof.

Using the solution of the inhomogeneous discrete state equation, Eq. 5.41, the first n number of output can be obtained as (here $\Delta t = 1$ now, for simplicity of notation)

$$y_1 = \mathbf{H\Phi}x_0 + \mathbf{H\Gamma}u_0 \qquad (5.63)$$

$$y_2 = \mathbf{H\Phi}^2 x_0 + \mathbf{H\Phi\Gamma}u_0 + \mathbf{H\Gamma}u_1$$

$$\vdots$$

$$y_n = \mathbf{H\Phi}^n x_0 + \mathbf{H\Phi}^{n-1}\mathbf{\Gamma}u_0 + \ldots + \mathbf{H\Gamma}u_{n-1}.$$

Defining

$$\mathbf{U}_n = [u_0, u_1, \ldots, u_{n-1}]^T \qquad (5.64)$$

and

$$\mathbf{Y}_n = [y_1, y_2, \ldots, y_n]^T \qquad (5.65)$$

Eq. 5.63 can be written as (Szöllősi-Nagy, 1987)

$$\mathbf{Y}_n = \begin{bmatrix} \mathbf{H\Phi} \\ \mathbf{H\Phi}^2 \\ \vdots \\ \mathbf{H\Phi}^n \end{bmatrix} x_0 + \begin{bmatrix} h_1 & 0 & \cdots & 0 \\ h_2 & h_1 & \ddots & \vdots \\ \vdots & \ddots & \ddots & 0 \\ h_n & h_{n-1} & \cdots & h_1 \end{bmatrix} \mathbf{U}_n \qquad (5.66)$$

where $h_j = \mathbf{H\Phi}^{j-1}\mathbf{\Gamma}$ is the jth ordinate ($j = 1, 2, \ldots, n$) of the discrete unit-pulse response of DLCM (see Eq. 5.45), which can be explicitly calculated by Eq. 5.47. The $n \times n$ quadratic matrix multiplying x_0 from the left is the observability matrix, Θ_n, of the discrete cascade. According to Theorem 17, Θ_n is observable; thus it is not singular [i.e. rank(Θ_n) = n], which means that it has an inverse. The initial condition, x_0, can be expressed from Eq. 5.66 by inverting it and denoting the matrix that contains the unit-pulse response values by $\mathbf{H}_n$, as

$$\mathbf{x}_0 = \mathbf{\Theta}_n^{-1}(\mathbf{Y}_n - \mathbf{H}_n\mathbf{U}_n). \tag{5.67}$$

The $\mathbf{H}_n\mathbf{U}_n$ vector's elements are the discrete convolutions (see Eq. 5.43) that yield the first n number of outputs provided the system is relaxed initially. This way the

$$\mathbf{e}_n = \mathbf{Y}_n - \mathbf{H}_n\mathbf{U}_n \tag{5.68}$$

vector reflects the effect of the initial condition. If $\mathbf{e}_n = \mathbf{0}$, then $\mathbf{Y}_n = \mathbf{H}_n\mathbf{U}_n$, which can only be if $\mathbf{x}_0 = \mathbf{0}$, i.e. the system was relaxed initially. By linearly transforming $\mathbf{e}_n$ with the help of the observability matrix, the unsteady initial condition is obtained. This is formulated in the following:

Theorem 18:   The initial state, $\mathbf{x}_0$, of the $\mathbf{\Sigma}_{DLCM}(\Delta t) = (\mathbf{\Phi}(\Delta t), \mathbf{\Gamma}(\Delta t), \mathbf{H})$ discrete cascade can be calculated unambiguously from the $[u_0, u_1, \ldots, u_{n-1}]^T$ and $[y_1, y_2, \ldots, y_n]^T$ input–output value pairs as (Szöllősi-Nagy, 1987)

$$\mathbf{x}_0 = \mathbf{\Theta}_n^{-1}\mathbf{e}_n. \tag{5.69}$$

Here $\mathbf{\Theta}_n$ is the discrete cascade's nonsingular observability matrix, described by Eq. 5.62, and

$$\mathbf{e}_n = \begin{bmatrix} y_1 - h_1 u_0 \\ y_2 - (h_2 u_0 + h_1 u_1) \\ \vdots \\ y_n - \sum_{j=0}^{n-1} h_{n-j} u_j \end{bmatrix} \tag{5.70}$$

where h_j ($j = 1, 2, \ldots, n$) is the *jth* ordinate (Eq. 5.47) of the discrete unit-pulse response of DLCM.

Note 5.21:   The initial condition, $\mathbf{x}_0$, is determined from the outputs (and the inputs that generate them) at time $t = 1, 2, \ldots, n$, via "backward" calculations. It is not by chance that the observability matrix plays a crucial role in the process, since it is this matrix that determines, through its definition, if such calculations are viable or not. If a system is not observable, then its observability matrix is singular; consequently, the initial state cannot be determined. Theorem 18 gives the algorithm as well.

Note 5.22:   The structure of the observability matrix does not show any particular feature that would help with analytical determination of its inverse. The inverse of the observability matrix must be obtained numerically. This should not pose a problem since the order of the cascade is usually very low for practical applications ($n < 5$).

Note 5.23:   When $n = 1$, the scalar initial condition, x_0, is obtained from the (u_0, y_1) input–output data pairs

$$\Theta_1 = ke^{-k}$$

from which the inverse is

$$\Theta_1^{-1} = \frac{1}{k}e^k.$$

The first value, h_1, of the unit pulse response is (from Eq. 5.47)

$$h_1 = 1 - e^{-k}.$$

This way the initial condition, x_0, becomes

$$x_0 = \frac{1}{k}e^k[y_1 - (1 - e^{-k})u_0]. \tag{5.71}$$

Theorem 18 must be true for the steady flow case as well, since no restrictions were made in the derivation of Eq. 5.69. The output equals the input, u_s, in a steady state. From Lemma 3, the steady state can be expressed (see Eq. 5.61) as $\mathbf{x}_s = k^{-1}\mathbf{i}u_s$, where $\mathbf{i} = [1, 1, \dots, 1]^T$. Eq. 5.67 in a steady state becomes

$$\mathbf{x}_s = k^{-1}\mathbf{i}u_s = \Theta_n^{-1}(\mathbf{i}u_s - \mathbf{H}_n\mathbf{i}u_s)$$

which after rearrangement yields

$$\mathbf{i} = \frac{1}{k}\Theta_n\mathbf{i} + \mathbf{H}_n\mathbf{i}. \tag{i}$$

By writing out the last row of Eq. (i), the following is obtained:

Corollary 6:   The system matrices, $\Phi(\Delta t)$, $\Gamma(\Delta t)$ (provided $k > 0$, $\Delta t > 0$), and $\mathbf{H}$ of the discrete cascade $\Sigma_{DLCM}(\Delta t)$, satisfy the

$$1 = \mathbf{i}_n\Phi^n(\Delta t)\mathbf{i} + \sum_{j=0}^{n-1}\mathbf{H}\Phi^j(\Delta t)\Gamma(\Delta t) \tag{5.72}$$

equation, where $\mathbf{i}_n = [0, 0, \dots, 1]$.

The practical importance of Eq. 5.72 is that it connects the system matrices in a way that helps check the correctness of the discrete cascade computer algorithm easily.

Note 5.24:   The second term of the right-hand-side of Eq. 5.72 is the unit-step response value (see Eq. 5.54) at $t = n\Delta t$. Considering this, the following can be written

$$1 - g_n(\Delta t) = \mathbf{i}_n \boldsymbol{\Phi}^n(\Delta t)\mathbf{i} \tag{5.73}$$

that, from Eq. 5.59, yields, if $n \longrightarrow \infty$

$$\lim_{n \to \infty} [1 - g_n(\Delta t)] = 0$$

from which it follows that

$$\lim_{n \to \infty} \boldsymbol{\Phi}^n(\Delta t) = 0. \tag{5.74}$$

This means that the elements of the state-transition matrix tend to zero with time, as was mentioned in the proof of Corollary 1. Eq. 5.74 follows from the theorem (Forsythe and Moler, 1967) that says for any $\mathbf{x}$ vector and $n \longrightarrow \infty$

$$\boldsymbol{\Phi}^n(\Delta t)\mathbf{x} \longrightarrow \mathbf{0} \tag{5.75}$$

but only if all eigenvalues of $\boldsymbol{\Phi}(\Delta t)$ have magnitudes less than unity. This property will be exploited below when studying the asymptotic behavior of the forecasts.

Note 5.25:   Eq. 5.69 is valid for every deterministic, discrete, linear, time-invariant dynamic system. However, Corollary 6 is valid for the DLCM only because of the specified structure of the state-input relationship. Hostetter (1982) recommends a recursive spectral analysis approach for the initial condition determination, while Sehitoglu (1982a,b) couples an identification technique, based on output errors, to the Ljapunov method, and proves that the initial condition estimations are satisfactory even with noisy data. The previously discussed recursive forecasting algorithm is purely deterministic, and so any corrupting noise in the data and the model will be dealt with in a stochastic submodel (described in Chapter 8), coupled with the deterministic model part. This way, application of the above referenced, computationally complex algorithms will be omitted.

5.6 DETERMINISTIC PREDICTION OF THE DISCRETE CASCADE OUTPUT AND ITS ASYMPTOTIC BEHAVIOR

Deterministic prediction of the state variable is given by Eq. 5.37, provided the state and input variables at time t are known, and that the input is zero at time $t + i\Delta t$. The conditional prediction of lead-time $i\Delta t$, $i \geqslant 1$, of the output is obtained, using Eqs. 5.23 and 5.46, as

$$
\begin{aligned}
y_{t+i\Delta t|t} &= \mathbf{H}\mathbf{x}_{t+i\Delta t|t} \\
&= \mathbf{H}\mathbf{\Phi}[(i-1)\Delta t]\mathbf{x}_{t+\Delta t|t} \\
&= \mathbf{H}\mathbf{\Phi}[(i-1)\Delta t][\mathbf{\Phi}(\Delta t)\mathbf{x}_t + \mathbf{\Gamma}(\Delta t)u_t] \\
&= \mathbf{H}\mathbf{\Phi}^i(\Delta t)\mathbf{x}_t + h_i(\Delta t)u_t. \tag{5.76}
\end{aligned}
$$

The first term on the right-hand-side of Eq. 5.76 tends to zero (Eq. 5.74) as $i \longrightarrow \infty$, and so does the second term (Eq. 5.49). This is trivial, since when there is no inflow, the reach slowly empties at an exponential rate, as indicated by the elements of the state-transition matrix. Falling discharges will not immediately follow the cessation of inflow, as is observed in Fig. 5.9, because the inflow at time t will have an effect on the storage up to the mean delay time of the reach: nK.

Assuming that $u_{t+i\Delta t} \equiv u_t$, $i \geq 1$, the output becomes

$$
y_{t+\Delta t|t} = \mathbf{H}\mathbf{\Phi}(\Delta t)\mathbf{x}_t + \mathbf{H}\mathbf{\Gamma}(\Delta t)u_t
$$

$$
y_{t+2\Delta t|t} = \mathbf{H}\mathbf{\Phi}^2(\Delta t)\mathbf{x}_t + \mathbf{H}\mathbf{\Phi}(\Delta t)\mathbf{\Gamma}(\Delta t)u_t + \mathbf{H}\mathbf{\Gamma}(\Delta t)u_t
$$

$$
\vdots
$$

$$
y_{t+i\Delta t|t} = \mathbf{H}\mathbf{\Phi}^i(\Delta t)\mathbf{x}_t + [\mathbf{H}\sum_{j=0}^{i-1}\mathbf{\Phi}^j(\Delta t)\mathbf{\Gamma}(\Delta t)]u_t \tag{5.77}
$$

which, from Eq. 5.54, can be written as

$$
y_{t+i\Delta t|t} = \mathbf{H}\mathbf{\Phi}^i(\Delta t)\mathbf{x}_t + g_i(\Delta t)u_t.
$$

In the limit, when $i \longrightarrow \infty$, the first term again tends to zero, while the second term, according to Eq. 5.59, tends to the steady input, u_t.

The above are summarized in the following:

Theorem 19: Asymptotic deterministic prediction of the output of the $\mathbf{\Sigma}_{DLCM}(\Delta t)$ discrete cascade is

$$
\lim_{i \longrightarrow \infty} y_{t+i\Delta t|t} = u_t \tag{5.78}
$$

provided

$$
u_{t+i\Delta t} = u_t, \quad \forall i \in (1, 2, \ldots). \tag{5.79}
$$

Note 5.26:   Eq. 5.79 follows from its continuity.

If for time $t + i\Delta t$, $i > 0$, the conditional forecasts of the input, $u_{t+i\Delta t|t}$ are available, which can be the outputs of DLCM from an upper stream-reach, the conditional forecasts can be written as

$$y_{t+i\Delta t|t} = \mathbf{H}\boldsymbol{\Phi}(i\Delta t)\mathbf{x}_t + \sum_{j=0}^{i-1} h_{i-j}(\Delta t)u_{t+j\Delta t|t} \tag{5.80}$$

which follows from the inhomogeneous solution of the state equation, (Eq. 5.41). For $j = 0$, $u_{t|t} = u_t$.

The above equation is valid for all discrete, linear, time-invariant SISO systems. From Eqs. 5.75 and 5.76 it follows that the effect of the initial condition on the predicted output reduces with time. That, however, does not diminish the importance of knowing the initial state, since it can be derived for any arbitrary time analytically, and so the recursive prediction can be started at any time. As a consequence, there is no need to start the model from a steady or near-steady state, as may be the case with a full dynamic wave model.

5.7 THE INVERSE OF PREDICTION: INPUT DETECTION

With DLCM one can quickly and accurately determine the inputs. Solution of this problem, known as *input detection* (Dooge, 1973), is not known for continuous hydrologic models. The reason for this is the difficulty of determining the initial condition of the system. As a result, input detection "has been widely ignored" in hydrology, which is an identification problem, and so it is "substantially more difficult than the problem of output prediction" (Dooge, 1973).

Note 5.27:   The problem of input detection can be found in the operation of flood-control reservoirs, where the outflow of the reservoir has to be regulated in a way that assures certain criteria are met concerning flow farther down the river (e.g. the maximum and minimum flow rates stay within a predefined interval). The same problem occurs in estimating effective precipitation distribution and time series from observed streamflow and in the estimation of missing upstream flow values using observed downstream flow values.

Thus, the task of input detection is an inverse problem: the input of a dynamic hydrologic system must be specified that results in an observed or prescribed output.

For the solution, let's assume that the parameters, n and k, of $\Sigma_{DLCM}(1)$, as well as the initial state, $\mathbf{x}_0$, and the output for $\tau = 1, 2, \ldots, t + 1 > n$ are known. Here it is shown that this information is sufficient to determine the input, u_t, of time t.

As before,

$$y_{t+1} = \mathbf{H}\mathbf{x}_{t+1} \tag{5.81}$$
$$= \mathbf{H}\mathbf{\Phi}\mathbf{x}_t + \mathbf{H}\mathbf{\Gamma}u_t$$
$$= \mathbf{H}\mathbf{\Phi}\mathbf{x}_t + h_1 u_t$$

where h_1 by definition is

$$h_1 = \mathbf{H}\mathbf{\Gamma} = 1 - e^{-k}\sum_{i=0}^{n-1}\frac{k^i}{i!}$$

which is the first ordinate (always positive) of the discrete cascade's unit-pulse response (see Eqs. 5.46 and 5.47) with the $\Delta t = 1$ choice. From Eq. 5.81 above (Szöllősi-Nagy, 1987)

$$\widehat{u}_t = \frac{1}{h_1}(y_{t+1} - \mathbf{H}\mathbf{\Phi}\mathbf{x}_t) \tag{5.82}$$

which for $t = 0, 1, 2, ..$ gives a recursive procedure for determining the input, where the hat denotes that it is an estimate and not a measured value. The recursion starts at $t = 0$, for which $\mathbf{x}_0$ is needed. For the estimation of the latter (see Eq. 5.67), the first n values of input–output are required. Input detection really starts at $t = n$, since up until $t = n - 1$, the inputs must be known for the calculation of $\mathbf{x}_0$. Consequently, the states can also be calculated recursively, once $\mathbf{x}_0$ has been estimated, as (Szöllősi-Nagy, 1987)

$$\mathbf{x}_n = \mathbf{\Phi}\mathbf{x}_{n-1} + \mathbf{\Gamma}u_{n-1} \tag{5.83}$$

from which $\mathbf{x}_n$, plus the observed output, y_{n+1}, yield $\widehat{u}_n$ via Eq. 5.82. As can be seen, the recursion consists of two steps: (1) calculation of the state at a given time (t) from the preceding state ($t - 1$); and (2) calculation of input at the given time (t) from the state at the same time (t) plus observed output at time $t + 1$.

Note 5.28:   When $u_t = 0$, then the output at $t + 1$ is

$$y_{t+1} = \mathbf{H}\mathbf{\Phi}\mathbf{x}_t = \mathbf{H}\mathbf{\Phi}(\mathbf{\Phi}\mathbf{x}_{t-1} + \mathbf{\Gamma}u_{t-1}) = \mathbf{H}\mathbf{\Phi}^2\mathbf{x}_{t-1} + \mathbf{H}\mathbf{\Phi}\mathbf{\Gamma}u_{t-1}$$

that can be considered as a conditional deterministic prediction of the output at $t + 1$ from information at $t - 1$. This way the error

$$\varepsilon_{t+1} = y_{t+1} - y_{t+1|t-1}$$

is used to detecting the input, u_t.

Input-detection
for $n = 1$

Example 5.3: When $n = 1$, x_0 can be calculated using Eq. 5.71, and $h_1 = 1 - e^{-k}$. Substituting these into Eq. 5.81, an identity for u_0 is obtained. From estimated x_0 and observed u_0, x_1 can be calculated via the state equation (Eq. 5.83), and so the $\widehat{u}_1$ input becomes, with the help of x_1 and observed y_2 output

$$\widehat{u}_1 = \frac{1}{1 - e^{-k}} (y_2 - ke^{-k}x_1). \tag{i}$$

Using the output equation, $y_t = Hx_t$, and inverting it yields, upon substitution into Eq. (i),

$$\widehat{u}_1 = \frac{1}{1 - e^{-k}} (y_2 - e^{-k}y_1).$$

With the help of x_1 and $\widehat{u}_1$, x_2 is calculable, and for $t = 2$ the following is similarly obtained

$$\widehat{u}_2 = \frac{1}{1 - e^{-k}} (y_3 - e^{-k}y_2)$$

and so on for every t. Fig. 5.10 illustrates the result of the above input detection for a cross-section of the Danube at Budapest, Hungary. DLCM was optimized for deterministic forecasting of the streamflow at Baja, about 200 km downstream from Budapest (Fig. 5.11), with observed flow-rate values at Budapest. The optimized parameters were: $n = 1$, $k = 0.6d^{-1}$. Flow measurements were taken daily, so $\Delta t = 1d$.

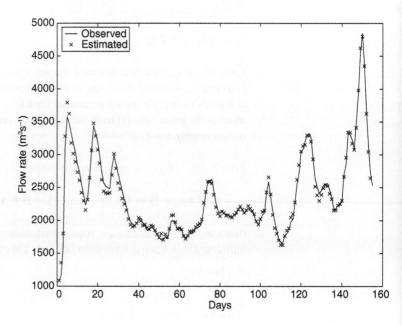

Figure 5.10. Input detection for the Danube at Budapest using observed discharge values at Baja, 200 km downstream.

Figure 5.11. Gauging station
locations.

The following summarizes the steps of input detection with $\Sigma_{DLCM}(1)$:

Algorithm 1:   *Step 1.* Determination of the initial state, x_0, through Eq. 5.69, using specified model parameters (n and k) and observed input-output values, $(u_0, u_1, \ldots, u_{n-1})$ and $(y_1, y_2, \ldots, y_n)$. Let $t \triangleq n$. *Step 2.* Calculation of state: $x_t = \Phi x_{t-1} + \Gamma u_{t-1}$. *Step 3.* Reading in y_{t+1}. *Step 4.* Calculation of input with Eq. 5.82. *Step 5.* Back to Step 2 with $t \triangleq t + 1$.

Note 5.29:   A prerequisite of the algorithm is observability of the system. The above algorithm is valid for any observable discrete linear dynamic system.

As was mentioned before, flood waves flatten out as they travel along the stream channel, which makes variance of streamflow at a downstream section generally smaller than at an upstream section, provided there is no tributary in between. The same is true with predictions and input detections: the latter always have higher variance than the former.

The Streeter-Phelps model

Example 5.4:   Here the discrete state space formulation of the continuous Streeter-Phelps model is discussed. The model describes changes in the water quality of a river and, due to its simplicity, it has become widely popular in the field, as it is still able to give meaningful and elegant answers to practical problems. The model assumes that the water quality of a river can be characterized by the dynamic interrelationship between the biochemical oxygen demand (BOD) and the dissolved oxygen (DO). Further, it assumes a first-order reaction kinetic for the BOD

$$\frac{dB(t)}{dt} = -K_r B(t)$$

where $B(t)$ is the BOD concentration in mg/l and K_r is the BOD removal or decay coefficient in day^{-1}. From continuity

$$\frac{dD(t)}{dt} = -K_a D(t) - K_r B(t) + K_a D_s$$

where $D(t)$ is the DO concentration in mg/l, K_a is the re-aeration co-efficient in day^{-1}, and D_s is the saturation level of the dissolved oxygen. Defining the state variables as $x_1(t) = B(t)$ and $x_2(t) = D(t) - D_s$, respectively, the latter being known as oxygen deficit and having direct physical meaning, the state equation of the Streeter-Phelps model becomes

$$\dot{\mathbf{x}}(t) = \mathbf{F}\mathbf{x}(t)$$

with

$$\mathbf{F} = \begin{bmatrix} -K_r & 0 \\ -K_r & -K_a \end{bmatrix}$$

considered to be constant. One of the objectives of setting up a water quality model is to control the water quality to achieve a desirable level of quality. The water quality of a river might, for example, be controlled by, among other things, treatment plants and artificial aeration facilities located along the stream. We define the control vector as $\mathbf{u}(t) = [u_1(t), u_2(t)]^T$, where $u_1(t)$ is for the control of effluent dumping from the sewage treatment plant and $u_2(t)$ is for the artificial aeration carried out. The first control might mean, say, the operation rule of a retention reservoir receiving the effluent of the treatment plant; the second control is the timing schedule of the aeration facilities. Thus the process model becomes

$$\dot{\mathbf{x}}(t) = \mathbf{F}\mathbf{x}(t) + \mathbf{G}\mathbf{u}(t)$$

with

$$\mathbf{G} = \begin{bmatrix} 1 & 0 \\ 0 & -1 \end{bmatrix}.$$

The minus sign refers to the fact that the more intensive the artificial aeration the less the oxygen deficit, and vice versa. Now, we are ready to derive a discrete model of the continuous process given above. According to Eq. 5.16, the state transition between the two time instants t and $t + 1$, is defined by the

$$\mathbf{\Phi}(t + 1, t) = e^{\mathbf{F}}$$

matrix exponential for a time-invariant system. Since the eigenvalues of $\mathbf{F}$ are negative, $\lambda_1 = -K_r$ and $\lambda_2 = -K_a$, the system is stable. Applying the well-known Sylvester expansion theorem, the one-step state-transition matrix is obtained as

$$\mathbf{\Phi}(t+1,t) = \begin{bmatrix} e^{-K_r} & 0 \\ \frac{-K_r}{K_a-K_r}\left(e^{-K_r} - e^{-K_a}\right) & e^{-K_a} \end{bmatrix} \tag{i}$$

provided $K_a \neq K_r$. As for the determination of the input-transition matrix $\mathbf{\Gamma}(t)$, Eq. 5.17 is evaluated with $\mathbf{G}$ above and, due to the special structure of the latter, it is equal to $\mathbf{\Phi}$, except that the lower-right matrix element has an additional negative sign. This way the state equation results as

$$\mathbf{x}(t+1) = \mathbf{\Phi}\mathbf{x}(t) + \mathbf{\Gamma}\mathbf{u}(t).$$

As far as the output of the system is concerned, the situation is that the evaluation of the BOD concentration usually requires several days in a laboratory, so for real-time control policies only DO measurements are available. That is

$$y(t) = \mathbf{H}\mathbf{x}(t)$$

where $\mathbf{H} = [0, \ 1]$. The system thus is specified by the $\mathbf{\Sigma} = (\mathbf{\Phi}, \mathbf{\Gamma}, \mathbf{H})$ triplet. The dynamics of this water quality control system is shown in Fig. 5.12.

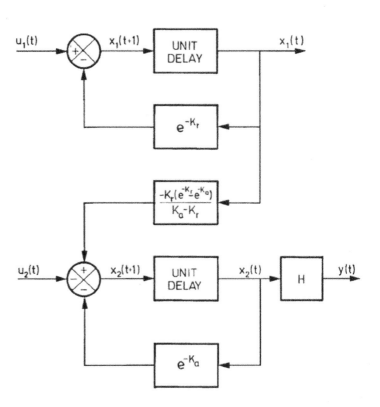

Figure 5.12. The dynamics of the discrete-time water quality control system.

Unit-pulse response
of the Streeter-Phelps
model

Example 5.5:   Next we determine the unit-pulse response (h) of the
above water quality control system. Let us assume that the system at
$t = 0$ is initially relaxed, i.e. $\mathbf{x}(0) = 0$ (or it is transformed into a relaxed
system from an equilibrium state $[\mathbf{x}'(0) = \mathbf{x}^*]$ as $\mathbf{x}(0) = \mathbf{x}'(0) - x^*$. From
Eq. 5.46 and with $\Delta t = 1$, the unit-pulse response function values for
$i \geq 1$ can be obtained as

$$\mathbf{h}_i \quad = \quad \mathbf{H}\mathbf{\Phi}(i-1)\mathbf{\Gamma}$$

$$= \quad [0,\ 1] \begin{bmatrix} e^{-(i-1)K_r} & 0 \\ \dfrac{-K_r}{K_a - K_r}\left(e^{-(i-1)K_r} - e^{-(i-1)K_a}\right) & e^{-(i-1)K_a} \end{bmatrix}$$

$$\begin{bmatrix} e^{-K_r} & 0 \\ \dfrac{-K_r}{K_a - K_r}\left(e^{-K_r} - e^{-K_a}\right) & -e^{-K_a} \end{bmatrix}$$

which yields

$$\mathbf{h}_i = \begin{bmatrix} \dfrac{-K_r}{K_a - K_r}\left[\left(e^{(1-i)K_r} - e^{(1-i)K_a}\right)e^{-K_r} + \left(e^{-K_r} - e^{-K_a}\right)e^{(1-i)K_a}\right] \\ -e^{-iK_a} \end{bmatrix}^T .$$

Observability of the
Streeter-Phelps model

Example 5.6:   In order to estimate nonmeasured state-variables, it is
important to determine whether the system is observable. If the system
is not observable then the internal state variables cannot be determined
or estimated. Let us examine whether the Streeter-Phelps water quality
model is observable, i.e. can we determine the BOD values from DO
measurements and under what conditions. For notational simplicity let
Eq. (i) of Example 5.4 be

$$\mathbf{\Phi} = \begin{bmatrix} \Phi_{11} & 0 \\ \Phi_{21} & \Phi_{22} \end{bmatrix} .$$

Since $\mathbf{H} = [0,\ 1]$, from Eq. A1.18, the observability matrix for $n = 2$
becomes

$$\mathbf{\Theta} = \left[\mathbf{H}^T \vdots \mathbf{\Phi}^T \mathbf{H}^T\right] = \begin{bmatrix} 0 & \Phi_{21} \\ 1 & \Phi_{22} \end{bmatrix} ,$$

which has a rank of two, or is invertible only if $\Phi_{21} \neq 0$, i.e. if

$$\frac{-K_r}{K_a - K_r}\left(e^{-K_r} - e^{-K_a}\right) \neq 0.$$

First, consider the case when $K_r \neq K_a$. Obviously

$$0 < \left|\frac{1}{K_a - K_r}\right| < c_1 < \infty$$

and

$$0 < \left| e^{-K_r} - e^{-K_a} \right| < c_2 < \infty .$$

Consequently,

$$0 < \left| \frac{1}{K_a - K_r} \right| \left| e^{-K_r} - e^{-K_a} \right| < c_1 c_2 < \infty$$

therefore if $K_r \neq K_a$, then $\Phi_{21} \neq 0$. Now consider the possibility that $K_r = K_a$. Then

$$\frac{-K_r e^{-K_r} \left[1 - e^{-(K_a - K_r)} \right]}{K_a - K_r} = \frac{0}{0}$$

which is an indeterminate form. Thus let $K_a - K_r = K$ and consider

$$\lim_{K \to 0} K_r e^{-K_r} \left[\frac{1 - e^K}{K} \right]$$

for which the L'Hospital rule yields

$$\lim_{K \to 0} K_r e^{-K_r} \left[\frac{-e^K}{1} \right] = -K_r e^{-K_r} \neq 0.$$

Thus if K_r and K_a are nonzero and bounded, the observability matrix is nonsingular, consequently the system is completely observable. To gain more insight to the notion of observability, let us make a change in the water quality system, namely, assume that only BOD data are available for control. Then in this new system the output matrix is $\mathbf{H}_* = [1, 0]$ and the observability matrix becomes

$$\Phi = \begin{bmatrix} 1 & \Phi_{11} \\ 0 & 0 \end{bmatrix}$$

which is of rank one, i.e. this system is unobservable.

This chapter derived the deterministic model-component of the forecasting model and described its properties. It was shown that a trivial discretization of the continuous system is not adequate, i.e. discrete coincidence, continuity and transitivity are all violated. DLCM, on the other hand, was shown to be unconditionally conservative, discretely coincident, and transitive, provided $\Delta t \longrightarrow 0$ for the last property. It was demonstrated that DLCMs with different sampling intervals are related to each other through a linear transformation. It was proven that the discrete linear kinematic wave and DLCM are equivalent. DLCM was also shown to be observable, and its initial condition always calculable.

System characteristics of the DLCM, which play a role in predicting the state and output variables and determine their asymptotic behaviors in the pulse-data framework, were identified. The inverse task of prediction was discussed, and an algorithm was given for input detection of the DLCM.

EXERCISES

5.1. Using the definition of the incomplete gamma-function, show that for $n = 1$ Eq. 5.20 is true.

5.2. Explain why $\Phi(\Delta t)$ is a matrix and why $\Gamma(\Delta t)$ is only a vector for the zero lateral inflow case. Note that later in the book each storage element may receive input not only from the one upstream but laterally as well, representing a tributary and/or groundwater contribution. Can you explain why we see the impulse responses in a decreasing order in each row of the state-transition matrix? What is the physical explanation of it? *Hint:* think about additivity of a linear system.

5.3. Why does the state-transition matrix contain the impulse responses but the input-transition vector does not? What are the elements in the latter and why are they ordered as they are?

5.4. Write up the homogeneous ordinary differential equation of the storage for $n = 1$ and solve it by separation of the variables keeping in mind that the solution is the impulse response written for the storage. Show that Eq. 5.37 is indeed equal to this homogeneous solution for any positive i.

5.5. Show that the unit pulse response, $h_{i\Delta t}$, of DLCM sums to unity as $i \to \infty$, and that $h_{i\Delta t} \to 0$, as $i \to \infty$.

5.6. Write out the discrete unit-pulse response function, $h_{i\Delta t}$, for $n = 1$.

5.7. Show that $h_{i\Delta t}$, $i = 1, 2, \ldots$ is discretely coincident with the continuous unit pulse response function for $n = 1$.

5.8. Starting with Eq. 5.55 show that the unit-step response is $1 - e^{-ki\Delta t}$ for $n = 1$, $i > 0$.

5.9. Show that Eq. 5.56 really follows from Eq. 5.55 for $t = \Delta t$.

5.10. For $n = 1$ and $i = 1, 2$ demonstrate Eq. 5.57.

5.11. Check Eq. 5.72 for $n = 1$ and $n = 2$ by hand.

5.12. What is the estimate of x_0 with $n = 1$, $\Delta t = 1$, $k = 0.6$, $u_0 = 1084$, $y_1 = 1286$?

5.13. What is the estimate of $\mathbf{x}_0$ with $n = 2$, $\Delta t = 1$, $k = 1.2$ now, if in addition to the measured in- and outflows in the previous example $u_1 = 1153$, $y_2 = 1318$? What is the y_3 prediction?

5.14. Choose a stream section of your liking with no tributaries. With trial and error (or with an optimization technique of your preference) calibrate n and k for a given period using discharge measurements taken at regular intervals (days for example). With each (n, k) pair use Eq. 5.72 to make sure you wrote up $\Phi(\Delta t)$ and $\Gamma(\Delta t)$ correctly. Then you have a choice: either estimate the initial state with each parameter pair, or just start with a relaxed system. In the latter case, you will need to discard the first couple of values for performance statistics calculations (e.g. least-square sum) on which your calibration is based, coming from the so-called "spin-up" period that allows the model to reach the correct state variable value. With the calibrated parameters, perform an input detection as well for a few time-intervals. Once you can accomplish all this, you have mastered application of the Discrete Linear Cascade Model, at least in a pulse-data system framework, after which modifying it for an LI-data system should be straightforward. You can find sample MATLAB scripts in Appendix II to assist you with your own coding.

CHAPTER 6

The Linear Interpolation (LI) Data System Approach

So far, within the pulse-data system framework, it has been assumed that the value of the continuous variable, sampled at discrete time-instants, remains constant between subsequent samplings. This assumption was convenient in deriving the input-transition matrix of the discrete linear dynamic system (Eq. 5.10), since the input, $u(\tau)$, being a constant over the time interval, $[t, t + \Delta t)$, could be brought outside the integral in the definition of the input-transition matrix, $\boldsymbol{\Gamma}$ (Eq. 5.12). In case of flow-routing, it is more realistic to assume that the input variable does not stay constant over the sampling interval, Δt, but rather, that it changes linearly. As the size of Δt decreases, a linear-change approach becomes ever more accurate, since the nonlinear terms in the Taylor-expansion vanish ever faster. Assuming a linear change in a continuous variable's value over the sampling interval results in the *linear interpolation* or *LI-data system* approach.

6.1 FORMULATION OF THE DISCRETE CASCADE IN THE LI-DATA SYSTEM FRAMEWORK

The discrete state equation (Eq. 5.10) has to be re-evaluated in the new data framework, as

$$
\mathbf{x}(t + \Delta t) = \boldsymbol{\Phi}(t + \Delta t, t)\mathbf{x}(t) + \int_{t}^{t+\Delta t} \boldsymbol{\Phi}(t + \Delta t, \tau)\mathbf{G}(\tau)u(\tau)d\tau
$$

$$
= \boldsymbol{\Phi}(\Delta t)\mathbf{x}(t) + \int_{t}^{t+\Delta t} \boldsymbol{\Phi}(t + \Delta t - \tau)\mathbf{G}u(\tau)d\tau \tag{6.1}
$$

where above it made use of time-invariance and the fact that $\mathbf{G}$ is a constant vector for the continuous KMN cascade. Note that the state-transition matrix remains the same as in the pulse-data system case, but the second term of Eq. 6.1 is different from the one in Eq. 5.10.

Let's evaluate the term

$$\mathbf{\Gamma}^s(\Delta t) \triangleq \int\limits_{t}^{t+\Delta t} \mathbf{\Phi}(t + \Delta t - \tau)\mathbf{G}u(\tau)d\tau$$

within the LI-data system. For clarity, the evaluation will be performed for the ith component, $\gamma_{i,1}^s(\Delta t)$, of the $n \times 1$ vector, $\mathbf{\Gamma}^s(\Delta t)$. Accounting for the linear change in $u(\tau)$ over Δt, and recalling that $\mathbf{G} = [1, 0, \ldots, 0]^T$, the ith component of $\mathbf{\Gamma}^s(\Delta t)$ can be written as

$$
\begin{aligned}
\gamma_{i,1}^s(\Delta t) &= \int\limits_{t}^{t+\Delta t} \Phi_{i,1}(t + \Delta t - \tau)u(\tau)d\tau \\
&= \int\limits_{t}^{t+\Delta t} \Phi_{i,1}(t + \Delta t - \tau)[u(t) + \frac{u(t + \Delta t) - u(t)}{\Delta t}(\tau - t)]d\tau \\
&= \int\limits_{t}^{t+\Delta t} [\Phi_{i,1}(t + \Delta t - \tau)u(t) + \Phi_{i,1}(t + \Delta t - \tau)\frac{u(t + \Delta t) - u(t)}{\Delta t}\tau \\
&\quad -\Phi_{i,1}(t + \Delta t - \tau)\frac{u(t + \Delta t) - u(t)}{\Delta t}t]d\tau.
\end{aligned}
\tag{6.2}
$$

Performing a change of variables as $\xi = k(t + \Delta t - \tau)$, the first term on the right-hand-side of Eq. 6.2 becomes

$$\frac{u(t)}{k(i-1)!} \int\limits_{0}^{\Delta tk} \frac{\xi^{(i-1)}}{e^\xi}d\xi = \frac{u(t)}{k(i-1)!}\Gamma(i, \Delta tk) = \frac{1}{k}\frac{\Gamma(i, \Delta tk)}{\Gamma(i)}u(t)$$

where the $\Gamma(i, \Delta tk)$ term is the so-called incomplete gamma function. Here the identity, $(i - 1)! = \Gamma(i)$, was used again for integers. Similarly, the third term (including the minus sign) of Eq. 6.2 can be expressed as

$$-\frac{u(t + \Delta t) - u(t)}{\Delta tk}\frac{\Gamma(i, \Delta tk)}{\Gamma(i)}t.$$

The second term requires a few more steps since both $\Phi_{i,1}(t + \Delta t - \tau)$ and the τ multiplier depend on the integral variable. Performing the same change of variables, gives

$$\frac{u(t + \Delta t) - u(t)}{\Delta tk(i-1)!} \int\limits_{0}^{\Delta tk} \frac{\xi^{(i-1)}}{e^\xi}\left(t + \Delta t - \frac{1}{k}\xi\right)d\xi$$

$$= \frac{u(t + \Delta t) - u(t)}{\Delta t}\left[\frac{(t + \Delta t)\Gamma(i, \Delta tk)}{k\Gamma(i)} - \frac{1}{k^2(i-1)!}\int\limits_{0}^{\Delta tk}\frac{\xi^{(i-1)}}{e^\xi}\xi d\xi\right]$$

$$= \frac{u(t+\Delta t) - u(t)}{\Delta t} \left[\frac{(t+\Delta t)\Gamma(i,\Delta tk)}{k\Gamma(i)} - \frac{1}{k^2} \frac{\Gamma(i+1,\Delta tk)}{\Gamma(i)} \right]$$

$$= \frac{u(t+\Delta t) - u(t)}{\Delta t} \left[\frac{(t+\Delta t)\Gamma(i,\Delta tk)}{k\Gamma(i)} - \frac{1}{k^2} \frac{i\Gamma(i,\Delta tk) - (\Delta tk)^i e^{-\Delta tk}}{\Gamma(i)} \right]$$

where the algebraic identity $\Gamma(a+1,x) = a\Gamma(a,x) - x^a e^{-x}$ (Abramowitz and Stegun, 1965) was used. After combining all three terms, $\gamma_{i,1}^s$ becomes

$$\gamma_{i,1}^s(\Delta t) = \int_t^{t+\Delta t} \Phi_{i,1}(t+\Delta t - \tau)u(\tau)d\tau$$

$$= \frac{1}{k}\frac{\Gamma(i,\Delta tk)}{\Gamma(i)} \left\{ [1 + \Lambda_{i,1}(\Delta t)]u(t+\Delta t) - \Lambda_{i,1}(\Delta t)u(t) \right\}$$

(6.3)

where the $\Lambda_{i,1}(\Delta t)$ term is defined as

$$\Lambda_{i,1}(\Delta t) = \frac{(\Delta tk)^{i-1}e^{-\Delta tk}}{\Gamma(i,\Delta tk)} - \frac{i}{\Delta tk}.$$

(6.4)

When making the following additional definitions

$$\gamma_{i,1}^{s1}(\Delta t) \triangleq -\frac{1}{k}\frac{\Gamma(i,\Delta tk)}{\Gamma(i)}\Lambda_{i,1}(\Delta t)$$

(6.5)

$$\gamma_{i,1}^{s2}(\Delta t) \triangleq \frac{1}{k}\frac{\Gamma(i,\Delta tk)}{\Gamma(i)}[1 + \Lambda_{i,1}(\Delta t)]$$

(6.6)

Eq. 6.1 can finally be written in the LI-data system framework as (Szilágyi, 2003)

$$\mathbf{x}_{t+\Delta t} = \Phi(\Delta t)\mathbf{x}_t + \Gamma^{s1}(\Delta t)u_t + \Gamma^{s2}(\Delta t)u_{t+\Delta t}$$

(6.7)

(compare it with Eq. 5.15) where the first input-transition vector, $\Gamma^{s1}(\Delta t)$, is

$$\Gamma^{s1}(\Delta t) = \begin{bmatrix} \frac{1}{k}\frac{\Gamma(1,\Delta tk)}{\Gamma(1)}\left[\frac{1}{\Delta tk} - \frac{e^{-\Delta tk}}{\Gamma(1,\Delta tk)}\right] \\ \frac{1}{k}\frac{\Gamma(2,\Delta tk)}{\Gamma(2)}\left[\frac{2}{\Delta tk} - \frac{(\Delta tk)e^{-\Delta tk}}{\Gamma(2,\Delta tk)}\right] \\ \vdots \\ \frac{1}{k}\frac{\Gamma(n,\Delta tk)}{\Gamma(n)}\left[\frac{n}{\Delta tk} - \frac{(\Delta tk)^{n-1}e^{-\Delta tk}}{\Gamma(n,\Delta tk)}\right] \end{bmatrix}$$

(6.8)

and the second input-transition vector, $\mathbf{\Gamma}^{s2}(\Delta t)$, is defined as

$$
\mathbf{\Gamma}^{s2}(\Delta t) = \begin{bmatrix} \dfrac{1}{k}\dfrac{\Gamma(1,\Delta tk)}{\Gamma(1)}\left[1 + \dfrac{e^{-\Delta tk}}{\Gamma(1,\Delta tk)} - \dfrac{1}{\Delta tk}\right] \\[2ex] \dfrac{1}{k}\dfrac{\Gamma(2,\Delta tk)}{\Gamma(2)}\left[1 + \dfrac{(\Delta tk)e^{-\Delta tk}}{\Gamma(2,\Delta tk)} - \dfrac{2}{\Delta tk}\right] \\[2ex] \vdots \\[1ex] \dfrac{1}{k}\dfrac{\Gamma(n,\Delta tk)}{\Gamma(n)}\left[1 + \dfrac{(\Delta tk)^{n-1}e^{-\Delta tk}}{\Gamma(n,\Delta tk)} - \dfrac{n}{\Delta tk}\right] \end{bmatrix}. \tag{6.9}
$$

The two new input-transition vectors can be related to the input-transition vector of the pulse-data system model. Eq. 6.5 can be equally written as

$$
\gamma_{i,1}^{s1}(\Delta t) = \frac{i}{\Delta tk}\gamma_{i,1}(\Delta t) - \Phi_{i,1}(\Delta t) \tag{6.10}
$$

and similarly for Eq. 6.6

$$
\gamma_{i,1}^{s2}(\Delta t) = \left(1 - \frac{i}{\Delta tk}\right)\gamma_{i,1}(\Delta t) + \Phi_{i,1}(\Delta t) \tag{6.11}
$$

where $\gamma_{i,1}(\Delta t)$ is the ith component of the input-transition vector, $\mathbf{\Gamma}(\Delta t)$, of the pulse-data system (see Eq. 5.19). By defining the diagonal matrix, $\mathbf{D}(\Delta t)$, as $\mathbf{D}(\Delta t) \triangleq < 1/\Delta tk, \ldots, i/\Delta tk, \ldots, n/\Delta tk >$, the two input-transition matrices can be written as

$$
\begin{aligned}
\mathbf{\Gamma}^{s1}(\Delta t) &= \mathbf{D}(\Delta t)\mathbf{\Gamma}(\Delta t) - \mathbf{\Phi}(\Delta t)\mathbf{G} & (6.12)\\
\mathbf{\Gamma}^{s2}(\Delta t) &= [\mathbf{I} - \mathbf{D}(\Delta t)]\mathbf{\Gamma}(\Delta t) + \mathbf{\Phi}(\Delta t)\mathbf{G}. & (6.13)
\end{aligned}
$$

Note that now there are two inputs required in the state equation (Eq. 6.7). This is so because in the LI-data framework the input value changes linearly between samplings, and a first-order polynomial requires two parameters to be identified unambiguously.

Theorem 20: For pulsed data, the state equations are identical in the two data frameworks.

Proof: In the LI-data framework, input is represented by straight lines of different slopes between samplings. For a pulsed data this means that the two input values at t and $t + \Delta t$ must be the same in the LI-data framework to be consistent with the zero slope value of the pulsed data: $u(t + \Delta t) = u(t)$. Inserting this identity into Eq. 6.7 and using Eqs. 6.12 and 6.13, results in Eq. 5.15, which concludes the proof.

Corollary 7: The unit-pulse and unit-step responses are the same in both data systems.

Note 6.1: The unit-pulse responses may indeed be identical in the two data frameworks; however, in the LI-data system, the unit-pulse response loses its property of providing, through discrete convolution, the output of an originally relaxed discrete system. It is so because now the input is defined by two values instead of one, and now there is an infinite number of possibilities for the input's shape over the sampling interval due to the existence of infinite possible slope values. The possibility that the input could always be decomposed into a unit pulse as $u(\tau) = \alpha u_p(\tau)$, $t \leqslant \tau < t + \Delta t$, where α is an arbitrary positive number, no longer exists. As a consequence, the unit-pulse response has no particular significance in the LI-data framework.

A signal starting at a in t and linearly changing to reach b in $t + \Delta t$ can, however, be decomposed into the sum of two linear ramp functions: one that starts at a in t and reaches zero in $t + \Delta t$, and another one that starts at zero in t and reaches b in $t + \Delta t$. For these ramp functions, proportionality will be valid, i.e. the first one can be obtained as a times the unit linear ramp function with a negative slope, and the second one as b times the unit linear ramp function with a positive slope. By definition, a unit linear ramp function starts at unity and ends at zero (having a negative slope) Δt time later and vice versa, starts at zero and ends at unity (positive slope). Consequently, Eq. (6.8), when multiplied by k, is the response function to the negative-sloped unit ramp input and Eq. (6.9) is that of the positive-sloped one.

Theorem 21: The two discrete approaches, described by Eqs. 5.15 and 6.7, are equivalent with pulsed inputs.

Proof: The state equations of the two discrete systems are identical, provided the input is pulsed. Consequently, the two systems have identical output values at discrete time increments to identical pulsed inputs. This concludes the proof.

Although the two approaches are equivalent with pulsed inputs, it does not mean that the two give the same discrete output values to the same discrete input sequence, as demonstrated in Fig. 6.1. The reason is that the two approaches assume different behavior of the input signal between samplings. Fig. 6.1 demonstrates again that the discrete model, now within the LI-data system, is discretely coincident, which follows again from the state equation (Eq. 6.1) defining the state trajectory between two points in time separated by Δt.

Note that when making operational forecasts with Eq. 6.7, the input at time $t + \Delta t$ is not known yet; only a prediction of it may be available. In Fig. 6.1 these input values were taken to be known. Such modeling is called *simulation* and, trivially, it is always more accurate than forecasting.

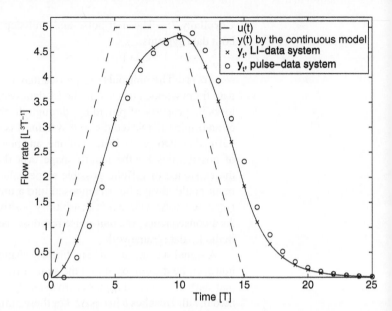

Figure 6.1. Outflow (y) of an initially relaxed system to hypothetical inflow (u), $n = 1$, $k = 0.5\ [T^{-1}]$, $\Delta t = 1[T]$.

Fig. 6.1 also demonstrates the advantage of an LI-data framework over a pulsed one. In the latter, the input at $t < \tau < t + \Delta t$ is always taken equal to the input at time t, due to the pulsed nature of the assumed input behavior between discrete samplings when making a prediction at time t. However, we can make any assumption about the input value at $t + \Delta t$ in the LI-data framework, which gives a significant additional flexibility and advantage in forecasting. This advantage is the clearest when reliable forecasts are available at the inflow cross-section of the given river reach.

Theorem 22: The discrete linear cascade, $\Sigma_{DLCM}(\Delta t)$, keeps its continuity (i.e. remains conservative) in the LI-data framework.

Proof: As it was shown earlier, if a system is conservative, then in a steady state the output equals the input. In a steady state, the input is constant, which means that the state equation of the LI-data framework reduces to that of the pulse-data system, for which continuity has already been proven. This concludes the proof.

Theorem 23: Convergence to transitivity improves in the LI-data framework.

Proof: As was shown with Theorem 9, the discrete cascade is not transitive in general because of the difference in the assumed discrete and continuous system responses between two consecutive discrete sampling instants. If it is true that the LI-data framework reduces this difference as $\Delta t \longrightarrow 0$, then the system must approach transitivity faster than in the pulse-data framework. But this is so, because from Eq. 6.1 the difference at $t + \tau$, $\tau < \Delta t$ in the LI-data framework is

$$y(t + \tau) - \hat{y}(t + \tau) = y(t + \tau) - \left(y_t + \frac{y_{t+\Delta t} - y_t}{\Delta t} \tau \right) \tag{i}$$

where the hat denotes the assumed value of the discrete signal between samplings in the actual data-framework. The same difference in the pulse-data framework is

$$y(t + \tau) - \hat{y}(t + \tau) = y(t + \tau) - y_t. \tag{ii}$$

Because of discrete coincidence, $y_t = y(t)$ and $y_{t+\Delta t} = y(t + \Delta t)$ can be substituted in the above equations. The square of Eq. (i) is indeed always smaller than that of Eq. (ii), because, when going through the calculations, the inequality

$$\left| \frac{dy(t)}{dt} \right| < 2 \left| \frac{dy(t)}{dt} \right| \tag{iii}$$

is obtained where it was considered that $\Delta t \longrightarrow 0$ when replacing the finite differences with the corresponding derivatives. This concludes the proof.

As in the pulse-data system, the question arises of how models with different equidistant sampling intervals relate to each other in the LI-data system framework. The state transition matrices are the same in both representations; thus the corresponding transformation matrix must remain the same as it was in the pulse-data system. The transformation matrices for the two new input-transition matrices can be obtained by inserting the new input-transition matrices into Eq. 5.29. The following transformation matrices of diagonal form are obtained:

$$\mathbf{T}_{\Gamma^{s1}}(\mu) = < \ldots, \frac{1}{\mu} \frac{i\Gamma(i, \mu\Delta tk) - (\mu\Delta tk)^i e^{-\mu\Delta tk}}{i\Gamma(i, \Delta tk) - (\Delta tk)^i e^{-\Delta tk}}, \ldots > \tag{6.14}$$

$$\mathbf{T}_{\Gamma^{s2}}(\mu) = < \ldots, \frac{1}{\mu} \frac{\Gamma(i, \mu\Delta tk)(\mu\Delta tk - i) + (\mu\Delta tk)^i e^{-\mu\Delta tk}}{\Gamma(i, \Delta tk)(\Delta tk - i) + (\Delta tk)^i e^{-\Delta tk}}, \ldots >$$
$$\tag{6.15}$$

where the terms shown are the ith components of the two diagonals.

The initial condition for predictions can be calculated similar to the pulse-data framework (Eq. 5.63) with the obvious distinction that now there are two inputs at every discrete time instant

$$y_1 = \mathbf{H\Phi x_0} + \mathbf{H\Gamma}^{s1}u_0 + \mathbf{H\Gamma}^{s2}u_1 \tag{6.16}$$

$$y_2 = \mathbf{H\Phi}^2\mathbf{x_0} + \mathbf{H\Phi}(\mathbf{\Gamma}^{s1}u_0 + \mathbf{\Gamma}^{s2}u_1) + \mathbf{H\Gamma}^{s1}u_1 + \mathbf{H\Gamma}^{s2}u_2$$

$$\vdots$$

$$y_n = \mathbf{H\Phi}^n\mathbf{x_0} + \mathbf{H\Phi}^{n-1}(\mathbf{\Gamma}^{s1}u_0 + \mathbf{\Gamma}^{s2}u_1) + ... + \mathbf{H}(\mathbf{\Gamma}^{s1}u_{n-1} + \mathbf{\Gamma}^{s2}u_n)$$

where for simplicity of writing, $\Delta t = 1$ was again assumed. Denoting

$$\mathbf{Y}_n = [y_1, \cdots, y_n]^T; \quad \mathbf{U}_n^{(1)} = [u_0, \cdots, u_{n-1}]^T; \quad \mathbf{U}_n^{(2)} = [u_1, \cdots, u_n]^T$$

together with

$$\mathbf{H}_n^{(1)} = \begin{bmatrix} \mathbf{H\Gamma}^{s1} & 0 & \cdots & 0 \\ \mathbf{H\Phi\Gamma}^{s1} & \mathbf{H\Gamma}^{s1} & \ddots & \vdots \\ \vdots & \ddots & \ddots & 0 \\ \mathbf{H\Phi}^{n-1}\mathbf{\Gamma}^{s1} & \cdots & \mathbf{H\Phi\Gamma}^{s1} & \mathbf{H\Gamma}^{s1} \end{bmatrix} \tag{6.17}$$

and

$$\mathbf{H}_n^{(2)} = \begin{bmatrix} \mathbf{H\Gamma}^{s2} & 0 & \cdots & 0 \\ \mathbf{H\Phi\Gamma}^{s2} & \mathbf{H\Gamma}^{s2} & \ddots & \vdots \\ \vdots & \ddots & \ddots & 0 \\ \mathbf{H\Phi}^{n-1}\mathbf{\Gamma}^{s2} & \cdots & \mathbf{H\Phi\Gamma}^{s2} & \mathbf{H\Gamma}^{s2} \end{bmatrix} \tag{6.18}$$

Eq. 6.16 can be written as

$$\mathbf{Y}_n = \mathbf{\Theta}_n\mathbf{x_0} + \mathbf{H}_n^{(1)}\mathbf{U}_n^{(1)} + \mathbf{H}_n^{(2)}\mathbf{U}_n^{(2)} \tag{6.19}$$

where $\mathbf{\Theta}_n$ is the same observability matrix of the discrete system as was defined in Eq. 5.66. Inverting the above equation yields

$$\mathbf{x_0} = \mathbf{\Theta}_n^{-1}[\mathbf{Y}_n - (\mathbf{H}_n^{(1)}\mathbf{U}_n^{(1)} + \mathbf{H}_n^{(2)}\mathbf{U}_n^{(2)})] = \mathbf{\Theta}_n^{-1}\mathbf{e}_n. \tag{6.20}$$

Note that even though the observability matrix is the same, the initial condition is different in the two data frameworks with the same observations, simply because the assumed system behavior between discrete observations is different in the two frameworks. The only exception is in a steady state, when the two input-transition matrices collapse to the input-transition matrix of the pulsed system, being the output (and input) of the system constant. Consequently, the system diagnostic equation, Eq. 5.72, remains in effect by writing out $\mathbf{\Gamma}(\Delta t)$ as $\mathbf{\Gamma}^{s1}(\Delta t) + \mathbf{\Gamma}^{s2}(\Delta t)$. As can be seen, $n+1$ input and n output values are needed for determining the initial condition of an *n*-order cascade in the LI-data framework, which means one extra piece of data in comparison with the pulse-data system.

Deterministic prediction derives from Eq. 6.16 as

$$y_{t+i\Delta t|t} = \mathbf{H}\mathbf{\Phi}^i(\Delta t)\mathbf{x}_t + \mathbf{H}\sum_{j=0}^{i-1}\mathbf{\Phi}^{i-1-j}(\Delta t)[\mathbf{\Gamma}^{s1}u_{t+j\Delta t|t} + \mathbf{\Gamma}^{s2}u_{t+(j+1)\Delta t|t}]$$

(6.21)

where $u_{t|t} = u_t$. Compare this equation with Eqs. 5.80 and 5.46. In both equations, forecasts for the upstream cross-section of the river are included in the prediction of the downstream flow, provided $i > 1$. An important difference exists for $i = 1$, i.e. for the one-step forecast. The LI-data system can incorporate upstream forecasts in the one-step prediction, while the pulse-data system cannot. Improvement in the one-step forecast affects multi-step forecasts, as evidenced by the forecast equation above. In nested conditions, when reliable one-step forecasts are available for the upstream cross-section, the LI-data system is expected to be better than the pulse-data system forecasts. This is demonstrated in the illustrations (from Szilágyi, 2003) below (Figs. 6.2 and 6.3), where simulation results are shown for Baja at the Danube, about 200 km downstream from Budapest, the upstream station, for arbitrary ($i \geqslant 1$) days of lead-time. The use of the words "simulation" and "multi-step lead-time" are compatible as long as the calculation of $y_{t+i\Delta t}$ ($i = 1, 2, \ldots$) starts with $\mathbf{x}_t$ in Eq. 6.21. Note that this way simulations can be considered as best-case scenarios of nested forecasts, i.e. the upstream forecasts are "perfectly on target"! Observe the "forecast" improvement at the peak values of the two largest floods of the period between the two data frameworks. Note that when perfect upstream forecasts are available then the forecasts do not deteriorate with increasing lead-time. Thus the one-day forecast has the same accuracy as the *i*-day ($i > 1$) forecast.

Naturally, when no forecasts are available (i.e. $u_{t+(j+1)\Delta t|t} = u_{t+j\Delta t|t} = u_t, j > 0$) for the upstream section, the two frameworks give the same result, since then the two input-transition matrices of the LI-data framework collapse to the input-transition matrix of the pulse-data system (Eq. 5.77).

There remains the discussion of input detection within the LI-data framework. With the help of Eq. 6.7 and $\Delta t = 1$ for simplicity of notations, the output at time t can be written as

$$y_{t+1} = \mathbf{H}\mathbf{\Phi}\mathbf{x}_t + \mathbf{H}(\mathbf{\Gamma}^{s1}u_t + \mathbf{\Gamma}^{s2}u_{t+1})$$

(6.22)

from which the detected input, $\widehat{u}$ becomes

$$\widehat{u}_{t+1} = \frac{1}{\mathbf{H}\mathbf{\Gamma}^{s2}}(y_{t+1} - \mathbf{H}\mathbf{\Phi}\mathbf{x}_t - \mathbf{H}\mathbf{\Gamma}^{s1}u_t).$$

(6.23)

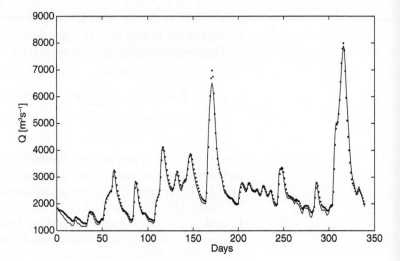

Figure 6.2. Measured and DLCM-simulated (dots) flow values (arbitrary i-day [$i \geqslant 1$] lead-time) of the Danube, Budapest – Baja. Pulse-data framework.

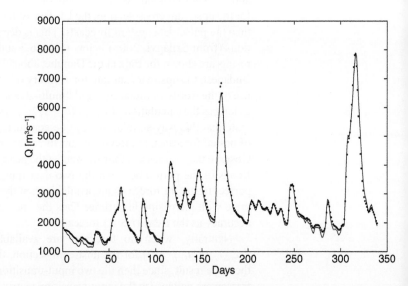

Figure 6.3. Measured and DLCM-simulated (dots) flow values (arbitrary i-day [$i \geqslant 1$] lead-time) of the Danube, Budapest – Baja. LI-data framework.

Note that the $\mathbf{H}\boldsymbol{\Gamma}^{s2}$ term is a scalar. With the help of the state equation written as

$$\mathbf{x}_t = \boldsymbol{\Phi}\mathbf{x}_{t-1} + \boldsymbol{\Gamma}^{s1}u_{t-1} + \boldsymbol{\Gamma}^{s2}u_t \tag{6.24}$$

the first n discrete states at times $t = 1, \cdots, n$ can be calculated, since for the estimation of the initial condition, $\mathbf{x}_0$, inputs at $t = 0, \cdots, n$ and outputs at $t = 1, \cdots, n$ must be known. The first detected input is at $t = n + 1$ for which all necessary variables are known in Eq. 6.23. With the detected input, Eq. 6.24 can be applied for $\mathbf{x}_{n+1}$, by which the input can be detected at time $t = n + 2$, and so on. The recursion, however, can become highly unstable since Eq. 6.23 includes the previously estimated input beside the measured output. This way input detection within the LI-data system framework has limited practical applicability.

6.2 DISCRETE STATE–SPACE APPROXIMATION OF THE CONTINUOUS KMN-CASCADE OF NONINTEGER STORAGE ELEMENTS

There seems to be one major difference between the continuous KMN-cascade and its state–space formulated version (either continuous or discrete in time) of it. Namely, the impulse–response function (Eq. 2.22) of the original cascade, when generalized, can take up noninteger values of n by simply replacing the factorial with the complete gamma function. In practical applications this feature can be advantageous.

Note 6.2: The complete gamma function, $\Gamma(n)$, is defined for all rational numbers, while the factorial is only defined for integer n values. In such cases $\Gamma(n) = (n - 1)!$, as known.

In the state–space approach there can only be an integer number of storage elements. However, the routing results obtained with Eq. 2.22 of noninteger n can still be approximated using the following considerations.

The impulse–response (Eq. 2.22) of a single storage element, when $n < 1$, is also given by Eq. 2.22 written as

$$h(t) = k\frac{(kt)^{n-1}}{\Gamma(n)}e^{-kt}. \tag{6.25}$$

In the state–space formulation a trivial choice for a constant storage coefficient when $x = n < 1$ can be $k_x = k/x$ (Szilágyi, 2006) since the mean storage time $K = k^{-1}$ is expected to be smaller for a fractional storage element than for a full one (i.e. when $n = 1$). With this constant coefficient approximation a fractional storage element will behave as a full one with a magnified k value. This observation also means that the uniform fractional n-cascade (i.e. when n is noninteger) of Eq. (6.25) can be represented in the state–space approach by replacing the last storage element in the cascade with an element whose storage coefficient is $k_x = k[n - int(n)]^{-1}$, where *int* designates the integer part of n. As a simplifying convention, the fractional element *must* always be the last one in the cascade, ensuring that only the last row of the system matrices are different to the case of a uniform cascade. Note that the order of the unequal storage elements is otherwise irrelevant since any ordering results in the same output due to linearity (Dooge and O'Kane, 2003, pp. 90).

The new $n^* \times n^*$ [where $n^* = int(n+1)$] system matrix, $\mathbf{F}$, will remain unchanged in its $int(n) \times int(n)$ dimension, but its last row/column will be changed

$$
\mathbf{F} = \begin{bmatrix}
-k & & & & & 0 \\
k & -k & & & & \\
& k & -k & & & \\
& & \ddots & \ddots & & \\
0 & & & k & -\dfrac{k}{x}
\end{bmatrix}
\tag{6.26}
$$

where $x = n - int(n)$. Determination of the new state-transition matrix, $\mathbf{\Phi}$, can be achieved by e.g. successive convolution. Note that unlike in the system matrix case, each element of the last row of $\mathbf{\Phi}$ will be different. Performing the matrix exponential in Eq. 4.9 for small values of n^* with the help of, e.g. the Maple software, it can deduced that the last row will contain the impulse–responses (divided by k_x) of nonuniform cascades of decreasing (by unity) dimension, similarly to the last row of $\mathbf{\Phi}$ in Eq. 4.9 that contains the impulse–responses (divided by k) of integer uniform n-cascades. Note that $\Phi_{n^*,n^*} = e^{-k_x t}$. It is sufficient to determine $\Phi_{n^*,1}$, as

$$
\Phi_{n^*,1} = \frac{1}{k_x} h(t) = \frac{1}{k_x} \int_0^t k \frac{(k\tau)^{n^*-2}}{(n^*-2)!} e^{-k\tau} k_x e^{-k_x(t-\tau)} d\tau
\tag{6.27}
$$

which, after some algebraic manipulation, yields

$$
\Phi_{n^*,1} = \frac{k(kt)^{n^*-2} e^{-k_x t}}{(n^*-2)!(k_x-k)} \cdot
$$
$$
\left\langle e^{(k_x-k)t} + \left\{ [(k-k_x)\,t]^{2-n^*} \left[(n^*-2)\Gamma(n^*-2,(k-k_x)t) - (n^*-2)! \right] \right\} \right\rangle,
$$
$$
n^* \geq 2; \quad k_x \neq k
\tag{6.28}
$$

Note that when $n^* = 2$ and $k_x \neq k$, there is a cascade of two unequal linear storage elements.

Similarly to the state-transition matrix, the first $int(n)$ elements of the input-transition vector will be the same as in the uniform cascade case. The last component of $\mathbf{\Gamma}(\Delta t)$ can be obtained, as before, through successive convolution

$$
\gamma_{n^*,1}(\Delta t) = \frac{1}{k_x} g(\Delta t) = \frac{1}{k_x} \int_t^{t+\Delta t} \left[1 - e^{-k\tau} \sum_{j=0}^{n^*-2} \frac{(k\tau)^j}{j!} \right] k_x e^{-k_x(t-\tau)} d\tau
$$
$$
\tag{6.29}
$$

which, after certain degree of algebraic manipulation, becomes

$$\gamma_{n^*,1}(\Delta t) = \frac{1 - e^{-k_x \Delta t}}{k_x} -$$

$$e^{-k_x \Delta t} \sum_{j=0}^{n^*-2} \left\langle \frac{(k\Delta t)^j}{(k_x - k)j!} \left\{ e^{(k_x - k)\Delta t} + \left[\frac{j\Gamma(j,(k - k_x)\Delta t) - j!}{[(k - k_x)\Delta t]^j} \right] \right\} \right\rangle$$

$$n^* \geq 2; \qquad k_x \neq k. \tag{6.30}$$

Eqs. 6.28 (with the $t = \Delta t$ substitution) and 6.30 form the state–space approximation of a uniform fractional n-cascade written in a pulse-data system framework (Szilágyi, 2006). The state-transition matrix is the same in both the pulse- and LI-data system frameworks, but not, however, the input-transition vector.

The input-transition vector, as before, separates into two vectors in the LI-data system approach, one, $\mathbf{\Gamma}^{s1}(\Delta t)$, that operates on $u(t)$ and another, $\mathbf{\Gamma}^{s2}(\Delta t)$, that acts on $u(t + \Delta t)$. Again, the first $int(n)$ elements of either input-transition vectors remain unchanged

$$\gamma_{i,1}^{s1}(\Delta t) = \frac{i}{k\Delta t}\gamma_{i+1,1}(\Delta t) \qquad i = 1, \cdots, int(n) \tag{6.31}$$

and

$$\gamma_{i,1}^{s2}(\Delta t) = \gamma_{i,1}(\Delta t) - \frac{i}{k\Delta t}\gamma_{i+1,1}(\Delta t) \qquad i = 1, \cdots, int(n) \tag{6.32}$$

respectively, where the definition of Eq. 5.19 was used.

Note 6.3: Eqs. 6.31 and 6.32 are the same as Eqs. 6.5 and 6.6, only written in a more succinct form.

As before, the last component of the input-transition vectors can be obtained through successive convolution. After some algebraic manipulations, the successive convolution yields the following expressions for the last component of the input-transition vectors

$$\gamma_{n^*,1}^{s2}(\Delta t) = \frac{e^{-k_x \Delta t}}{\Delta t} \left[\alpha - \beta - \frac{n^* - 1}{k}\gamma_{n^*+1,1}(\Delta t) \right] \tag{6.33}$$

and

$$\gamma_{n^*,1}^{s1}(\Delta t) = \gamma_{n^*,1}(\Delta t) - \gamma_{n^*,1}^{s2}(\Delta t) \tag{6.34}$$

where

$$\alpha = \frac{1 + e^{k_x \Delta t}(k_x \Delta t - 1)}{k_x^2} \tag{6.35}$$

$$\beta = \sum_{j=1}^{n^*-1} \left\langle \frac{k^{j-1}(\Delta t)^j}{(k_x - k)(j-1)!} \left\{ e^{(k_x - k)\Delta t} + \left[\frac{j\Gamma(j, (k - k_x)\Delta t) - j!}{[(k - k_x)\Delta t]^j} \right] \right\} \right\rangle \tag{6.36}$$

respectively.

Finally, in the output equation's **H** vector, k_x will replace k for the nonuniform n^*-cascade. Eqs. 6.28 (with the $t = \Delta t$ substitution) and Eqs. 6.31 through 6.36 with the corresponding $u(t + \Delta t)$ and $u(t)$ values specify the state–space approximation of a uniform fractional n-cascade written now in a LI-data system framework (Szilágyi, 2006).

Figs. 6.4 and 6.5 display the impulse, unit-step, and ramp response functions of the uniform fractional n-cascade and the state–space formulated, nonuniform, discrete, integer n^*-cascade, written in an LI-data system framework. The constant slope of the ramp function applied is 0.1.

It can be concluded that the closer the value of n to an integer, the better the fit becomes between the uniform, fractional n-cascade and its approximate, state–space formulated nonuniform, integer n^*-cascade. Naturally, when n is an integer the two models are discretely coincident. Similarly, the larger the integer part of n, the smaller the difference becomes between the two model outputs. As a consequence, the two models are expected to yield almost identical forecasts when n is relatively large and/or when its value is close to an integer.

The importance of considering a fractional uniform cascade (and thus its nonuniform state–space approximation) is highlighted by the observation that in many practical applications, using flowrate values, the value of n tends to remain small. This is so because for a given stream reach, represented by uniform linear storage elements, the mean storage delay time (also called residence or travel time), T, is nk^{-1}. As the value of n is increased (while keeping T constant), the response of the river reach becomes less and less diffusive. Observations of natural river channels with a gentle slope (i.e. less than 0.01%, characteristic of the Danube in Hungary) show a typically high degree of dispersion (i.e. the flood waves flatten out relatively fast), thus leading to small optimized n values. For small values of n, however, it makes a relatively large difference whether n may assume only integer values or is allowed to have noninteger values as well during the optimization process.

Finally, specifying the system matrices for a discrete nonuniform cascade approximating a continuous uniform cascade of noninteger number of storage elements is necessitated by the fact that the Discrete Linear Cascade Model is transitive only when $\Delta t \longrightarrow 0$. Transitivity for any Δt would allow for taking the discrete output of a uniform $(n-1)$-cascade and

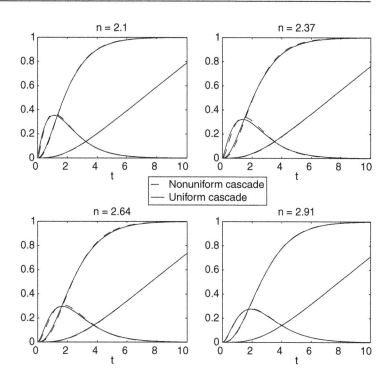

Figure 6.4. Impulse, unit-step, and ramp-response functions, $k = 1 \, [T^{-1}]$.

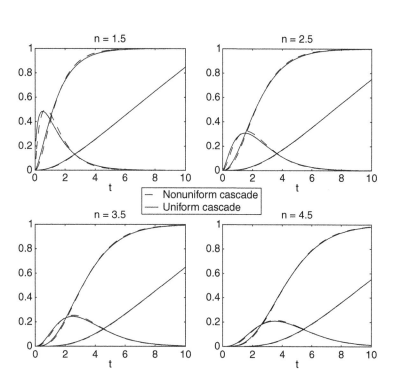

Figure 6.5. Impulse, unit-step, and ramp-response functions, $k = 1 \, [T^{-1}]$.

subsequently routing it through an additional storage element and obtain the same result as when performing the task in one single step, so there is no need to work out the system matrices of the nonuniform cascade case. This, however, is not so, simply, because the discrete model makes only assumptions on how the discretely observed input signal behaves between subsequent samples; consequently this assumed behavior of the input signal is not identical to that of the original continuous signal (Szilágyi, 2006). This way, two different signals enter the last storage element in the above example; thus the output must also be different between the two cases (i.e. one-step or two-step approach). Consequently, the output of a discrete nonuniform n-cascade *cannot* be replicated by simply employing a discrete uniform $(n-1)$-cascade first and routing its output additionally through another storage element (of different storage coefficient).

6.3 APPLICATION OF THE DISCRETE CASCADE FOR FLOW-ROUTING WITH UNKNOWN RATING CURVES

Below it is demonstrated how the KMN-cascade can be formulated for flow routing when there is no flow-rate information. For larger streams and for rivers, the primary source of flow information is in the form of stage measurements which are converted into instantaneous flow rates through the application of an established rating curve for the channel cross-section in question. A flow routing approach based solely on direct stage observations may prove useful when no rating curves are available or the rating curves are highly inaccurate.

The linear storage equation (Eq. 2.17) results if it is assumed that the exponent (α) is the same in the functional relationships between flow rate and stage as well as between water stored in a channel reach, $S(t)$, and stage

$$Q(t) \quad = \quad c_1[H(t) + a]^\alpha \tag{6.37}$$

$$S(t) \quad = \quad c_2[H(t) + a]^\alpha \tag{6.38}$$

where H $[L]$ is the measured value of the stage above or below datum, and c_1 $[L^{3-\alpha}T^{-1}]$, c_2 $[L^{3-\alpha}]$, and a $[L]$ are constants. Dividing Eq. 6.37 by Eq. 6.38 yields

$$Q(t) = \frac{c_1}{c_2}S(t) = kS(t). \tag{6.39}$$

Inserting Eqs. 6.37, 6.38, and 6.39 into the lumped continuity equation of the channel reach

$$\dot{S}(t) = Q^{(1)}(t) - Q^{(2)}(t) = Q^{(1)}(t) - kS(t) \tag{6.40}$$

results in

$$c_2\alpha[H^{(2)}(t) + a]^{\alpha-1}\frac{dH^{(2)}(t)}{dt} = -\frac{c_1}{c_2}c_2[H^{(2)}(t) + a]^{\alpha}$$

$$+ c_3[H^{(1)}(t) + b]^{\beta} \qquad (6.41)$$

where the superscripts 1 and 2 refer to the up- and downstream ends of the channel reach, and c_3 $[L^{3-\beta}T^{-1}]$, b $[L]$, and β are constants of the stage–discharge relationship of the upstream location. By rearranging Eq. 6.41,

$$\frac{dH^{(2)}(t)}{dt} = -\frac{c_1}{c_2\alpha}[H^{(2)}(t) + a] + \frac{c_3}{c_2\alpha}\frac{[H^{(1)}(t) + b]^{\beta}}{[H^{(2)}(t) + a]^{\alpha-1}} \qquad (6.42)$$

is obtained which shows that in general the future outflow rate of the reach is determined by a certain combination of in- and outflow rates through the last term of the right-hand-side of the equation. However, by assuming that both exponents are unity, Eq. 6.42 simplifies into

$$\frac{dH^{(2)}(t)}{dt} = -\frac{c_1}{c_2}H^{(2)}(t) + \frac{c_3}{c_2}H^{(1)}(t) + c_4 = -kH^{(2)}(t) + cH^{(1)}(t) + c_4 \qquad (6.43)$$

where $c = c_3/c_2$ $[T^{-1}]$, and c_4 $[LT^{-1}]$ are other constants. In comparison with Eq. 6.40 or 4.1, the constant multiplier of $H^{(1)}$ and an additional constant value now are of no concern because linearity assures that the output is proportional to any constant multiplier in the input values, and the presence of a constant input means only an additional constant value in the output values after an initial spin-up period. Because of the arbitrary reference points in the stage measurements of differing locations, routed upstream stage values have to be scaled up or down in any case to match the measured downstream stage values, thus the presence of a constant multiplier (and an additional constant) in the input stage values means no extra scaling. Consequently c and c_4 can be chosen arbitrarily. In this way Eq. 6.43 can be expressed as

$$\frac{dH^{(2)}(t)}{dt} = -kH^{(2)}(t) + H^{(1)}(t) \qquad (6.44)$$

which now is of the same form as Eq. 4.1 of the KMN-cascade when written for a single subreach. The reason why the required scaling is not typically a linear function stems from the general nonlinear shape of the actual rating curves, whereas in the derivation of Eq. 6.44 linear rating curves were used. The required scaling of routed to observed stage values can be achieved by the application of a polynomial curve fitting in the form of

$$\widehat{H^{(2)sc}}(t) = p_1\widehat{H^{(2)}}^m(t) + p_2\widehat{H^{(2)}}^{m-1}(t) + \cdots + p_m\widehat{H^{(2)}}(t) + p_{m+1}$$
(6.45)

where $\widehat{H^{(2)sc}}$ is the scaled, $\widehat{H^{(2)}}$ is the original model estimate of the downstream stage value, and the p_i-s $[L^{i-m}]$ are the constant coefficients of the polynomial of a predefined order m.

The discrete cascade over n serially connected subreaches can be written now as

$$\mathbf{H}_{t+\Delta t} = \mathbf{\Phi}(\Delta t)\mathbf{H}_t + \mathbf{\Gamma}^{s2}(\Delta t)H_{t+\Delta t}^{(1)} + \mathbf{\Gamma}^{s1}(\Delta t)H_t^{(1)}$$
(6.46)

where the vector $\mathbf{H}$ comprises the modeled stage values of the n subreaches, the $\mathbf{\Phi}(\Delta t)$, $\mathbf{\Gamma}_1(\Delta t)$ and $\mathbf{\Gamma}_2(\Delta t)$ are the same as in Eq. 6.7 before. The output equation now becomes

$$\widehat{H^{(2)}}(t) = [0, 0, \cdots, 1]\begin{bmatrix} H^{(1)}(t) \\ \vdots \\ H^{(n)}(t) \end{bmatrix}$$
(6.47)

the term on the left-hand-side being the input to Eq. 6.45. For channel reaches with tributaries, stages are routed separately between up- and downstream stations on the main channel and the upstream station of each tributary and the downstream station of the main channel due to linearity of the KMN-cascade, before inserting the $\widehat{H^{(2)j}}(t)$ $(j = 1, \cdots, T+1$, where T is the number of tributaries within the reach) values into Eq. 6.45. Then the p_i $(i = 1, \cdots, m)$ coefficients of the polynomial become vector-valued.

As a practical consideration, it can be mentioned that c_4 in Eq. 6.43 may need to be chosen different from zero in order to avoid negative values in the routing of stages when the upstream stage value can drop below datum.

Table 6.1 compares the performance of the present model with that of an operative forecasting model (discussed later) employed at the National Hydrological Forecasting Service of Hungary.

Here σ is the mean root-square error of forecasts, and a Nash-Sutcliffe-type efficiency coefficient is defined as

$$NSC = 100\left(1 - \frac{\sum_i(\widehat{H}_i - H_i)^2}{\sum_i(H_{i-1} - H_i)^2}\right) \quad [\%]$$
(6.48)

where $\widehat{H}_i$ is the predicted, and H_i the observed stage value on day i. The closer the NSC value is to 100% the better are the predictions. Note that

Table 6.1. Model performance statistics of the one-day ahead stage forecasts. The values in parentheses refer to the operative model (from Szilágyi et al., 2005).

	Optimization period (Jan. 1, 2000 – Dec. 31, 2001)
Budapest	$\sigma = 5.95$ (5.67) [cm], $NSC = 94.21$ (94.75) %
Dunaújváros	$\sigma = 6.58$ (8.42) [cm], $NSC = 92.15$ (87.14) %
Paks	$\sigma = 5.08$ (7.46) [cm], $NSC = 92.67$ (91.96) %
Baja	$\sigma = 6.92$ (5.68) [cm], $NSC = 91.75$ (94.43) %
Mohács	$\sigma = 5.28$ (5.49) [cm], $NSC = 94.34$ (93.90) %
Tokaj	$\sigma = 6.23$ (8.53) [cm], $NSC = 78.87$ (60.34) %
Makó	$\sigma = 12.02$ (11.85) [cm], $NSC = 66.79$ (67.72) %
	Verification period (Jan. 1, 2002 – Sep. 18, 2003) %
Budapest	$\sigma = 8.11$ (7.83) [cm], $NSC = 91.66$ (92.23) %
Dunaújváros	$\sigma = 8.59$ (9.88) [cm], $NSC = 89.13$ (85.75) %
Paks	$\sigma = 6.07$ (9.46) [cm], $NSC = 95.70$ (89.55) %
Baja	$\sigma = 7.69$ (7.87) [cm], $NSC = 91.81$ (91.45) %
Mohács	$\sigma = 6.16$ (6.72) [cm], $NSC = 93.79$ (92.61) %
Tokaj	$\sigma = 9.72$ (17.57) [cm], $NSC = 44.76$ (0) %
Makó	$\sigma = 9.36$ (10.85) [cm], $NSC = 64.01$ (51.49) %

the NSC value may be negative when the forecasts are worse than the naive prediction (see denominator), which takes the stage value of the actual day as the one-day forecast.

Overall, performance of the above model is very similar to that of the operative model. For certain stations (Budapest, Baja, and Makó) the operative model produces more accurate predictions than the recent model. This is what would normally be expected, since the operative model uses extra information (i.e. known rating curves) for flow routing. One plausible explanation of why the present model may perform better than the operative one for other stations (Dunaújváros, Paks, and Tokaj) can be that for those stations the rating curves may not be accurate enough or they may be outdated, i.e. they do not reflect correctly the channel and flow conditions of the modeled periods. Suboptimal parameter values (which could stem from a higher number of parameters to be optimized, i.e. 7 as opposed to 3) in the case of the operative model might also explain its underperformance, but it is unlikely knowing that parameter values of the operative model are updated each day using information from the previous 90 days (Szilágyi, 1992). Here it should be emphasized that the current model is not meant for replacing models that use measured rating-curve information. Whenever reliable rating curves are available, a flow-rate formulation should always be preferred over a stage formulation. However, an additional (on top of flow rates) flow routing using stages only, can detect inadequacies in the data required by the former. Naturally, when no information of rating curves is available, the proposed model (or its variant, such as a multilinear formulation) may easily be a proper candidate of a physically based model to apply.

6.4 DETECTING HISTORICAL CHANNEL FLOW CHANGES BY THE DISCRETE LINEAR CASCADE

Specific-stage diagrams of the Missouri River downstream of Omaha, Nebraska (Fig. 6.6), typically show increasing stage levels to fixed discharge values (Fig. 6.7), raising the spectre of an increased flood risk to the area and that despite the construction of a chain of major multi-purpose reservoirs upstream of Sioux City, Iowa.

DLCM was applied to model the flow over the 104-km long Nebraska City – Rulo section of the river in two distinct time periods: in the 1950s, before major river training works commenced to make the channel navigable for large barges, and in the 1990s, when such works had mostly been completed. Optimization resulted in $n = 3$ for both periods, but yielded $k = 5.7$ d^{-1} for the 1950s and $k = 4.3$ d^{-1} for the 1990s.

While previously it took about 0.53 day ($= n/k$) for a floodwave to travel the Nebraska City – Rulo distance, by the 1990s the same took about 0.7 day. These translate into mean celerities of 8.23 km/d and 6.21 km/d, respectively, a 25% slowing over time. Since flood celerity for a wide and relatively shallow rectangular channel can be approximated as $5d^{2/3}\sqrt{S_0}/3m$, where d is the mean channel depth, S_0 is the channel slope, and m is the Manning roughness coefficient of the channel, and where it could be ruled out that neither the mean channel depth nor the slope could decrease over time (mainly because of the continued dredging of the channel plus the intended purpose of wing-dyke construction, i.e. to concentrate and speed up the flow—to avoid sediment accretion—in

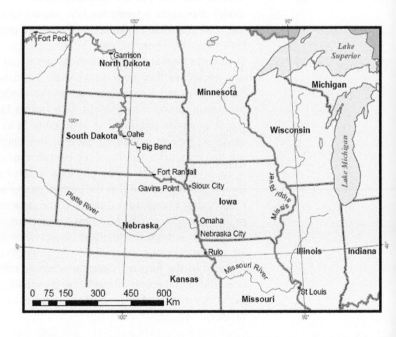

Figure 6.6. Location of the study reach (Nebraska City – Rulo) on the Missouri River. The dashes indicate dam locations on the river (from Szilágyi et al., 2008).

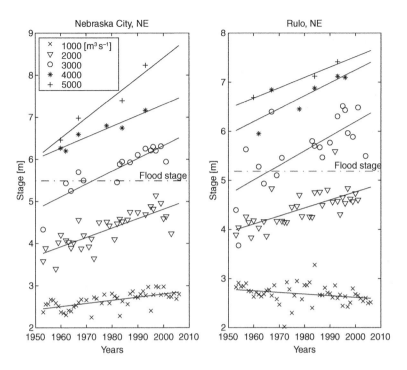

Figure 6.7. Specific-stage diagrams of the Missouri River at Nebraska City and Rulo, Nebraska, with linear trend functions fitted (from Szilágyi et al., 2008).

at least a narrow part of the channel and thus, to ensure a certain water depth for barge traffic), the celerity decrease could only be caused by a corresponding increase in the roughness coefficient. The latter most likely had been caused by a doubling of the number of wing-dykes within the reach, from 340 to about 660 over the period (Szilágyi et al., 2008).

The ease of application and minimal data requirement thus makes the DLCM a practical tool for streamflow analysis. It can also serve as a preliminary investigative tool for more advanced and detailed hydraulic approaches that typically require a data-rich environment and significantly greater development time.

This chapter derived the state equation of the discrete cascade in the LI-data framework. It was shown that the input-transition matrix of the pulse-data system decomposes into two matrices that transform the two inputs separated by Δt in the LI-data framework. It was also shown that discrete coincidence and continuity remains the same, while convergence to transitivity with $\Delta t \longrightarrow 0$ improves within the new framework. The two approaches were demonstrated to be identical with pulsed inputs, and so the unit-pulse and unit-step responses of the discrete cascade also become identical in the two frameworks. These characteristics, however, lose their significance in the LI-data system, because input can no longer be decomposed into unit-pulses in the new framework, since input is now defined by two separate values over each sampling interval. Estimation of the initial state and detection of inputs were demonstrated to be similar to the *pulse-data framework, but neither their calculation nor their estimated values are strictly identical in the two frameworks. It was also shown that predictions in the new data framework are expected to improve over the pulse-data approach in nested forecasts, when forecasts for the upstream cross-section of the stream are available. The discrete cascade model was next extended to allow for an approximation of a homogenous, fractional n-cascade response. A version of the discrete cascade that uses stage values rather than flow-rate ones was formulated for applications where no or just inaccurate information is available on the rating curves that transfer measured stage values into flow rates. Finally, it was demonstrated that the model can also serve as a practical, preliminary investigative tool for streamflow analysis before more sophisticated and expensive hydraulic approaches are used with significantly increased model-development time and data requirements.*

EXERCISES

6.1. Prove that a function linearly changing from a to b over Δt can always be decomposed as the sum of two linear ramp functions, one starting at a and reaching zero over Δt, and the other starting at zero and reaching b over the same time interval.

6.2. Derive Eq. (iii).

6.3. Show that the new transformation matrixes, Eqs. 6.14 and 6.15 are correct.

6.4. For $n = 2$, demonstrate that the output of the inhomogeneous cascade (i.e. $k_1 \neq k_2$) does not depend on the order of the storage elements.

6.5. What is the estimate of x_0 with $n = 1$, $\Delta t = 1$, $k = 0.6$, $u_0 = 1084$, $u_1 = 1153$, $y_1 = 1286$?

6.6. What is the estimate of $\mathbf{x}_0$ with $n = 2$, $\Delta t = 1$, $k = 1.2$ now, if in addition to the measured in- and outflows in the previous example $u_2 = 1580$, $y_2 = 1318$? What is the y_3 prediction?

6.7. Repeat Exercise 5-14 within the LI-data system framework.

CHAPTER 7

DLCM and Stream–Aquifer Interactions

The two examples below show how the Discrete Linear Cascade Model can be applied to account for the transfer of water between the channel and the adjacent aquifer. The first example describes the modifications needed in the state–space description of the DLCM to allow for considering bank storage and base-flow processes in flow routing. The second example shows how the actual rate of base-flow contribution to the channel can be estimated via the method of input detection, discussed in Chapter 5.

7.1 ACCOUNTING FOR STREAM–AQUIFER INTERACTIONS IN DLCM

It was shown previously that the discrete linear cascade model, $\Sigma_{DLCM}(\Delta t)$, is a special discretized form of the continuous linear kinematic wave equation that describes the translation of flood waves along the stream. Due to spatial discretization, the discrete cascade can account for the dispersion of the wave that causes it to flatten out as it travels. It has also been shown how tributary inflow can be incorporated into the model. However, there remains one important physical process that has not been considered yet, and that is flux exchange along the stream–aquifer interface. This exchange of water manifests itself as bank storage during flooding, which causes the peak of the flood wave to subside faster than it would otherwise due to dispersion only along its travel. The release of water from the banks after the flood in turn slows down the flow recession. Also, during prolonged periods without precipitation or snowmelt, the aquifer may supply groundwater to the stream solely responsible for maintaining its flow, which is referred to as *base flow*. These examples clearly show the need to include this exchange of water between stream and aquifer into our flow routing procedure.

   Flow, $q(t)$ $[L^2 T^{-1}]$, across the stream–aquifer interface (and over a unit length of the stream) can be described by Darcy's law under simplified conditions as (Hantush et al., 2002)

$$q(t) = P\kappa \frac{H(t) - h(0,t)}{b} \tag{7.1}$$

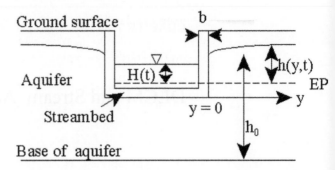

Figure 7.1. Schematic of the stream–aquifer system.

where P [L] is one half of the wetted perimeter of the stream, κ [LT^{-1}] is the saturated hydraulic conductivity of the streambed, b is the thickness of the streambed, $H(t)$ [L] is the water level in the stream above the reference, which can be an initial equilibrium position (EP) of the ground-water table, and $h(y, t)$ [L] is the elevation of the groundwater surface above the reference (Fig. 7.1). Of course, the total flow across the interface is twice that of Eq. 7.1 (provided conditions are similar) because the stream has two banks.

Before the above equation can be used with the discrete cascade, some further simplifying assumptions have to be made. These are: (a) the aquifer has a high enough diffusivity so that any water that crosses the streambed either from or to the aquifer would cause a change in $h(0, t)$ that is negligible compared to the mean saturated thickness of the aquifer; (b) changes in the groundwater surface elevations due to recharge and discharge are negligible to its overall height (h_0); and (c) the stored water volume, $x(t)$, in a stream reach (L) can be taken as proportional to $H(t)$. With these assumptions, Eq. 7.1 can be reformulated for a stream reach after taking account of both banks as

$$Q(t) = 2 \int_L \frac{P\kappa}{b} [H(t) - h(0, t)] \approx g[x(t) - x_0] \tag{7.2}$$

where $Q(t)$ now has a measurement unit of volume over time, and g [T^{-1}] can be conceptualized as the inverse of the mean delay time of storage (similar to k) in stream–aquifer interactions. Inserting Eq. 7.2 into the continuity equation (Eq. 4.1) of the storage element yields

$$\dot{x}(t) = u(t) - (k + g)x(t) + C_0 \tag{7.3}$$

where $C_0 = gx_0$ is a constant (Szilágyi, 2004a). For a cascade of storage elements, Eq. 7.3 becomes

$$
\dot{\mathbf{x}}(t) = \begin{bmatrix} -(k+g) & 0 & \cdots & 0 \\ k & -(k+g) & \ddots & \vdots \\ \vdots & \ddots & \ddots & 0 \\ 0 & \cdots & k & -(k+g) \end{bmatrix} \begin{bmatrix} x_1(t) \\ x_2(t) \\ \vdots \\ x_n(t) \end{bmatrix}
$$

$$
+ \begin{bmatrix} u(t) + C_0 \\ C_0 \\ \vdots \\ C_0 \end{bmatrix}
$$

$$
= \mathbf{Fx}(t) + \mathbf{u}(t) = \mathbf{Fx}(t) + \mathbf{Gu}(t) \tag{7.4}
$$

where there are four parameters: n, k, g, and C_0. The input-distribution matrix, $\mathbf{G}$, is just an $n \times n$ identity matrix: $\mathbf{G} = \mathbf{I}_n$, as was mentioned in Chapter 5. The state and input-transition matrices must be derived next.

The matrix-exponential of $\mathbf{F} = k(\mathbf{N}_n - \mathbf{I}_n) - g\mathbf{I}_n$ is

$$
e^{t\mathbf{F}} = e^{t[k(\mathbf{N}_n - \mathbf{I}_n) - g\mathbf{I}_n]} = e^{tk\mathbf{N}_n} e^{-t(k+g)\mathbf{I}_n} = e^{tk\mathbf{N}_n} < e^{-t(k+g)} > \tag{7.5}
$$

where the sharp brackets denote a diagonal matrix. The structure of the first term of the right-hand-side of Eq. 7.4 is the same as it was in Chapter 4, and so the discrete state-transition matrix becomes

$$
\boldsymbol{\Phi}(\Delta t) = \begin{bmatrix} e^{-\Delta t(k+g)} & 0 & \cdots & 0 \\ \Delta t k e^{-\Delta t(k+g)} & e^{-\Delta t(k+g)} & \ddots & \vdots \\ \vdots & \ddots & \ddots & 0 \\ \dfrac{(\Delta t k)^{n-1}}{(n-1)!} e^{-\Delta t(k+g)} & \cdots & \Delta t k e^{-\Delta t(k+g)} & e^{-\Delta t(k+g)} \end{bmatrix} \tag{7.6}
$$

which is similar to Eq. 5.18 except for the additional term of $-\Delta t g$ in the exponents. The discrete state equation can be obtained as

$$
\mathbf{x}(t + \Delta t) = \boldsymbol{\Phi}(\Delta t)\mathbf{x}(t) + \int_{t}^{t+\Delta t} \boldsymbol{\Phi}(t + \Delta t - \tau)\mathbf{Gu}(\tau)d\tau. \tag{7.7}
$$

For clarity of writing, the ith component of $\mathbf{x}(t)$ will only be considered below as in Eq. 6.2. Assuming that the system is relaxed at time t, gives

$$
x_{i,1}(t + \Delta t) = \int_{t}^{t+\Delta t} \Phi_{i,1}(t + \Delta t - \tau)u(\tau)d\tau
$$

$$
+ C_0 \int_{t}^{t+\Delta t} \sum_{j=1}^{i} \Phi_{i,j}(t + \Delta t - \tau)d\tau \tag{7.8}
$$

where the lower triangular property of the state-transition matrix was used. Performing a change of variables as $\xi = t + \Delta t - \tau$, the first term of the first integral (see Eq. 6.2) becomes (Szilágyi, 2003)

$$u(t) \frac{k^{i-1}}{(k+g)^i} \frac{\Gamma[i, \Delta t(k+g)]}{\Gamma(i)}$$

where the identity (Abramowitz and Stegun, 1965)

$$\int_0^\infty \frac{\xi^{(i-1)}}{e^{c\xi}} d\xi = \frac{1}{c^i} \Gamma(i) \tag{7.9}$$

was used. When the upper integral-limit is finite, X, Eq. 7.9 becomes (Szilágyi, 2004a)

$$\int_0^X \frac{\xi^{(i-1)}}{e^{c\xi}} d\xi = \frac{1}{c^i} \Gamma(i, cX). \tag{7.10}$$

The third term of the first integral similarly yields

$$\frac{k^{i-1}}{(k+g)^i} \frac{t[u(t) - u(t + \Delta t)]}{\Delta t} \frac{\Gamma[i, \Delta t(k+g)]}{\Gamma(i)}$$

whereas the second term becomes

$$\frac{k^{i-1}}{(i-1)!} \frac{u(t+\Delta t) - u(t)}{\Delta t} \int_0^{\Delta t} \frac{\xi^{(i-1)}}{e^{-(k+g)\xi}} (t + \Delta t - \xi) d\xi$$

$$= \frac{k^{i-1}}{(i-1)!} \frac{u(t+\Delta t) - u(t)}{\Delta t} \left[\frac{t+\Delta t}{(k+g)^i} \Gamma[i, \Delta t(k+g)] - \int_0^{\Delta t} \frac{\xi^i}{e^{-(k+g)\xi}} d\xi \right]$$

$$= \frac{k^{i-1}}{(i-1)!} \frac{u(t+\Delta t) - u(t)}{\Delta t} \frac{1}{(k+g)^i}$$

$$\times \left[(t+\Delta t)\Gamma[i, \Delta t(k+g)] - \frac{1}{k+g} \Gamma[i+1, \Delta t(k+g)] \right]$$

$$= \frac{k^{i-1}}{(k+g)^i} \frac{u(t+\Delta t) - u(t)}{\Delta t} \frac{1}{\Gamma(i)}$$

$$\times \left[(t+\Delta t)\Gamma[i, \Delta t(k+g)] - \frac{i\Gamma[i, \Delta t(k+g)]}{k+g} + [\Delta t(k+g)]^i e^{-\Delta t(k+g)} \right]$$

where again the following algebraic identity was applied (Abramowitz and Stegun, 1965)

$$\Gamma(i+1, z) = i\Gamma(i, z) - z^i e^{-z} \tag{7.11}$$

in addition to Eq. 7.10.

Combining all three terms, gives

$$
\int\limits_{t}^{t+\Delta t} \Phi_{i,1}(t + \Delta t - \tau)u(\tau)d\tau = \frac{k^{i-1}}{(k+g)^i} \frac{\Gamma[i, \Delta t(k+g)]}{\Gamma(i)}
$$

$$
= \left\{ [1 + \Lambda_{i,1}(\Delta t)]u(t + \Delta t) - \Lambda_{i,1}(\Delta t)u(t) \right\} \tag{7.12}
$$

with $\Lambda_{i,1}$ being

$$
\Lambda_{i,1}(\Delta t) = \frac{[\Delta t(k+g)]^{i-1}e^{-\Delta t(k+g)}}{\Gamma[i, \Delta t(k+g)]} - \frac{i}{\Delta t(k+g)}. \tag{7.13}
$$

Compare these with Eqs. 6.3 and 6.4.

The last term that remains to be evaluated is the second integral of Eq. 7.8. Since integration and summation commute,

$$
C_0 \int\limits_{t}^{t+\Delta t} \sum_{j=1}^{i} \Phi_{i,j}(t + \Delta t - \tau)d\tau = C_0 \sum_{j=1}^{i} \int\limits_{t}^{t+\Delta t} \Phi_{i,j}(t + \Delta t - \tau)d\tau
$$

$$
\tag{7.14}
$$

is obtained which, when $j = 1$, is the same as the first term of the first integral without the term $u(t)$ which is just a constant since t is set. Keeping track of j, Eq. 7.14 yields

$$
\Omega_{i,1}(\Delta t) = C_0 \sum_{j=1}^{i} \frac{k^{i-j}}{(k+g)^{i-j+1}} \frac{\Gamma[i - j + 1, \Delta t(k+g)]}{\Gamma(i - j + 1)} \tag{7.15}
$$

which is just a constant term.

Combining Eqs. 7.7, 7.11, 7.12, and 7.14 results in (Szilágyi, 2004a)

$$
x_{t+\Delta t} = \Phi(\Delta t)x_t + \Gamma_g^{s1}(\Delta t)u_t + \Gamma_g^{s2}(\Delta t)u_{t+\Delta t} + \Omega(\Delta t) \tag{7.16}
$$

where

$$
\Gamma_g^{s1}(\Delta t) = \begin{bmatrix} \dfrac{1}{(k+g)} \dfrac{\Gamma[1, \Delta t(k+g)]}{\Gamma(1)} \left[\dfrac{1}{\Delta t(k+g)} - \dfrac{e^{-\Delta t(k+g)}}{\Gamma[1, \Delta t(k+g)]} \right] \\[2ex] \dfrac{k}{(k+g)^2} \dfrac{\Gamma[2, \Delta t(k+g)]}{\Gamma(2)} \left[\dfrac{2}{\Delta t(k+g)} - \dfrac{\Delta t(k+g)e^{-\Delta t(k+g)}}{\Gamma[2, \Delta t(k+g)]} \right] \\[2ex] \vdots \\[2ex] \dfrac{k^{n-1}}{(k+g)^n} \dfrac{\Gamma[n, \Delta t(k+g)]}{\Gamma(n)} \left[\dfrac{n}{\Delta t(k+g)} - \dfrac{[\Delta t(k+g)]^{n-1}e^{-\Delta t(k+g)}}{\Gamma[n, \Delta t(k+g)]} \right] \end{bmatrix}
$$

$$
\tag{7.17}
$$

while

$$
\mathbf{\Gamma}_g^{s2}(\Delta t) =
\begin{bmatrix}
\dfrac{1}{(k+g)} \dfrac{\Gamma[1, \Delta t(k+g)]}{\Gamma(1)} \left[1 + \dfrac{e^{-\Delta t(k+g)}}{\Gamma[1, \Delta t(k+g)]} - \dfrac{1}{\Delta t(k+g)} \right] \\[3ex]
\dfrac{k}{(k+g)^2} \dfrac{\Gamma[2, \Delta t(k+g)]}{\Gamma(2)} \left[1 + \dfrac{\Delta t(k+g)e^{-\Delta t(k+g)}}{\Gamma[2, \Delta t(k+g)]} - \dfrac{2}{\Delta t(k+g)} \right] \\[3ex]
\vdots \\[2ex]
\dfrac{k^{n-1}}{(k+g)^n} \dfrac{\Gamma[n, \Delta t(k+g)]}{\Gamma(n)} \left[1 + \dfrac{[\Delta t(k+g)]^{n-1}e^{-\Delta t(k+g)}}{\Gamma[n, \Delta t(k+g)]} - \dfrac{n}{\Delta t(k+g)} \right]
\end{bmatrix}
\tag{7.18}
$$

and finally

$$
\mathbf{\Omega}(\Delta t) =
\begin{bmatrix}
C_0 \dfrac{1}{k+g} \dfrac{\Gamma[1, \Delta t(k+g)]}{\Gamma(1)} \\[3ex]
\vdots \\[2ex]
C_0 \displaystyle\sum_{j=1}^{n} \dfrac{k^{n-j}}{(k+g)^{n-j+1}} \dfrac{\Gamma[n-j+1, \Delta t(k+g)]}{\Gamma(n-j+1)}
\end{bmatrix}.
\tag{7.19}
$$

Note that the same relationship exists (Eqs. 6.12 and 6.13) between the two input-transition matrices as earlier, with the term $k+g$ replacing k in the diagonal matrix $\mathbf{D}$. Note also, that when g is zero, i.e. there is no interaction between the stream and the aquifer, then the system matrices become identical to Eqs. 6.8 and 6.9, and the $\mathbf{\Omega}(\Delta t)$ vector vanishes, since $C_0 = g x_0$.

Deterministic predictions are obtained similar to Eq. 6.21 as

$$
y_{t+i\Delta t|t} = \mathbf{H}\mathbf{\Phi}^i(\Delta t)\mathbf{x}_t + \mathbf{H}\sum_{j=0}^{i-1}\mathbf{\Phi}^{i-1-j}(\Delta t)
$$

$$
\times \left[\mathbf{\Gamma}_g^{s1} u_{t+j\Delta t|t} + \mathbf{\Gamma}_g^{s2} u_{t+(j+1)\Delta t|t} + \mathbf{\Omega}(\Delta t) \right].
\tag{7.20}
$$

Initial state calculation and input detection can be done as before with an extra term in Eqs. 6.19 and 6.20

$$
\mathbf{H}_n^{(3)} =
\begin{bmatrix}
\mathbf{H}\mathbf{\Omega} \\
\mathbf{H}\mathbf{\Phi}\mathbf{\Omega} \\
\vdots \\
\mathbf{H}\mathbf{\Phi}^{n-1}\mathbf{\Omega}
\end{bmatrix}
$$

and similarly, **Ω** in Eqs. 6.22 through 6.24. Note that the diagnostic equation (Eq. 5.72) is no longer valid because of the additional term, **Ω**.

The matrices $[\mathbf{T}_\Phi(\mu),\ \mathbf{T}_{\Gamma_g^{s1}}(\mu),\ \mathbf{T}_{\Gamma_g^{s2}}(\mu)$ and $\mathbf{T}_\Omega(\mu)]$ that transform states at Δt intervals to $\Delta t^* = \mu\Delta t$ intervals now can be obtained by substituting $k + g$ in place of k in the exponential term of Eq. 5.30, in place of all k terms in Eqs. 6.14 and 6.15, while the ith element of the $\mathbf{T}_\Omega(\mu)$ diagonal transformation matrix becomes

$$[T_\Omega(\mu)]_i = \frac{\displaystyle\sum_{j=1}^{i}\left\{ k^{j-1}(k+g)^{i-j}\Gamma\left[j,\mu\Delta t(k+g)\right]\prod_{m=j}^{i} m\right\}}{\displaystyle\sum_{j=1}^{i}\left\{ k^{j-1}(k+g)^{i-j}\Gamma\left[j,\Delta t(k+g)\right]\prod_{m=j}^{i} m\right\}}. \qquad (7.21)$$

Figs. 7.2 and 7.3 illustrate the importance of accounting for stream–aquifer interactions in streamflow forecasting.

Note that model simulation results improve not only under low-flow conditions, but during floods as well, as a result of accounting for stream–aquifer interactions in the discrete cascade. Notice also that due to groundwater discharge to the stream, flow rates may be higher downstream than the corresponding upstream flow values during low flow; that is why the model, without a stream–aquifer component, keeps undershooting those values, even though its parameter is optimized for best performance.

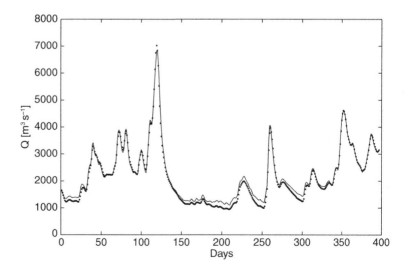

Figure 7.2. Measured and DLCM-simulated (dots) flow values (arbitrary i-day $[i \geqslant 1]$ lead-time) of the Danube, Budapest–Baja. LI-data framework, no stream–aquifer interactions included.

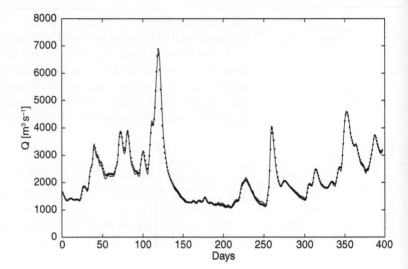

Figure 7.3. Measured and DLCM-simulated (dots) flow values (arbitrary i-day [$i \geqslant 1$] lead-time) of the Danube, Budapest–Baja. LI-data framework, stream–aquifer interactions included.

7.2 ASSESSING GROUNDWATER CONTRIBUTION TO THE CHANNEL VIA INPUT DETECTION

When only estimating the groundwater-discharge time series to the channel, $q(t)$, is of interest, it can be obtained as a simple input detection problem (Szilágyi et al., 2006) of lateral inflow (see Fig. 5.5) without resorting to the previously described augmentation of the transition matrixes. The state equation now, using a LI-data system approach, can be written as

$$\mathbf{x}_{t+\Delta t} = \mathbf{\Phi}(\Delta t)\mathbf{x}_t + \mathbf{\Gamma}^{s1}(\Delta t)u_t + \mathbf{\Gamma}^{s2}(\Delta t)u_{t+\Delta t} + \boldsymbol{\omega}(\Delta t)q_t \qquad (7.22)$$

where the new, additional $n \times 1$ input-transition vector's ith component becomes

$$\omega_{i,1} = \sum_{j=1}^{i} \gamma_j \qquad (7.23)$$

with γ_j given in Eq. 5.19.

Note 7.1: $\omega_{i,1}$ is the sum of the terms in the ith row of $\mathbf{\Gamma}(\Delta t)$ of Eq. 5.25. It is so because now the lateral inputs to each storage element are assumed to be equal and constant during each time increment.

After rearrangement of Eq. 7.22 combined with the measurement equation, the scalar-valued groundwater discharge to the channel, q_t, can be expressed (similar to Eq. 6.23) as

$$q_t = \frac{1}{\mathbf{H}\omega} \left(y_{t+\Delta t} - \mathbf{H}\boldsymbol{\Phi}\mathbf{x}_t - \mathbf{H}\boldsymbol{\Gamma}^{s1} u_t - \mathbf{H}\boldsymbol{\Gamma}^{s2} u_{t+\Delta t} \right) \qquad (7.24)$$

which is the desired groundwater contribution to the channel section between the up- and downstream gauging stations.

Input—in this case groundwater discharge to the channel—detection can be started in a period when the groundwater contribution to the channel is negligible (typically around the mean flow rate) in order to have the initial state estimated as accurately as possible using only the in- and outflow rates of the reach, since the groundwater contribution to the channel cannot typically be measured, and thus cannot be included in the initial state estimation procedure. Provided the parameters of the discrete cascade (i.e. n and k) have already been obtained, the first $n + 1$ inflow and n outflow values are used to estimate the initial state, $\mathbf{x}_0$, as described in Chapter 6. From Eq. 7.24, the first detected groundwater discharge to the channel is at $t = (n+1)\Delta t$. Note that the first inflow value is at $t = 0$. With the resulting q_t estimate, Eq. 7.22 is then updated, which in turn yields an updated state-variable vector to estimate the next groundwater-inflow value with Eq. 7.24 again.

Fig. 7.4 illustrates the resulting time series of the estimated groundwater contribution to the channel of the Danube between Budapest and Dunaföldvár (Fig. 5.11). The original groundwater-discharge estimates have been smoothed by a running average of five days (in both the forward and backward directions, in order to preserve the phase).

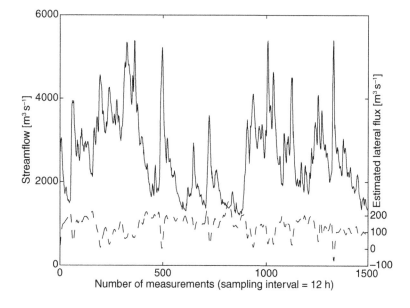

Figure 7.4. Stream discharge of the Danube at Dunaföldvár, January 1, 1995–January 19, 1997. Estimated groundwater discharge to the channel between Budapest and Dunaföldvár.

As expected, the groundwater discharge to the stream increases as the stage of the river falls, and decreases, as the stage increases. When the stage increase is abrupt, as seen near the end of the time period, the groundwater flow direction may reverse (negative values of the estimated lateral flux) and water flows from the channel to the adjacent aquifer lead to temporary bank storage.

In this chapter the discrete cascade, $\Sigma_{DLCM}(\Delta t)$ was expanded to include possible flux exchanges between the stream and the aquifer within the LI-data framework. Accounting for these interactions improves model accuracy not only during baseflow conditions but during flood events as well, since the model can now accommodate bank storage during floods and groundwater supply to the stream during low-flow periods. In an input-detection mode, the original discrete cascade, when formulated for lateral inflows, can also be used to estimate the time series of the groundwater discharge to the stream.

EXERCISES

7.1. Check for the correctness of the transformation matrix given by Eq. 7.21.

CHAPTER 8

Handling of Model Error:
The Deterministic–Stochastic Model
and Its Prediction Updating

Predictions $(\hat{y}_t)$ are rarely perfect, they contain varying degrees of error

$$\varepsilon_t = y_t - \hat{y}_t. \tag{8.1}$$

The error sequence may contain information that can improve future forecasts through *error updating*. Error updating is based on the model of errors and its predictions.

The most simple error model is called *sequential correction* (Bartha, 1970). It assumes that the model error of the actual forecast of lead-time Δt will be the same as it was the last time. The error correction this way becomes

$$\Delta y^*_{t+\Delta t} = \varepsilon_t \tag{8.2}$$

by which the updated forecast of lead-time Δt is

$$\hat{y}^*_{t+\Delta t|t} = y_{t+\Delta t|t} + \Delta y^*_{t+\Delta t} \tag{8.3}$$

where $y_{t+\Delta t|t}$ is the conditional deterministic prediction of output. This error updating is recursive but considers the error sequence to be static. An error updating that considers the dynamics in the error sequence is, however, much preferable over a static approach (Andjelić and Szöllősi-Nagy, 1980). The task now is to formulate a stochastic model for the errors defined by Eq. 8.1, and to update the deterministic model forecasts recursively.

8.1 A STOCHASTIC MODEL OF FORECAST ERRORS

Eqs. 5.15, 5.23, 6.7, and 7.16 specify the recursive deterministic predictions of the discrete cascade. Due to model and measurement uncertainties, these predictions may contain errors that are autocorrelated (see Fig. 7.3 for an illustration). A forecasting model works optimally, if the forecast-error time series/sequence form a *Gaussian white noise* (GWN) (Gelb, 1974).

Note 8.1:   A white noise sequence in discrete time contains values that are independent of each other (i.e. the values are truly random) and so they are unpredictable. The autocorrelation function of a white noise is a spike, which is unity at the origin and zero otherwise. The more the auto-correlation function of the forecast errors differs from this spike function, the more information it contains that can be harnessed by a stochastic model. Note here that the one-step forecast error is often called *residual* or *innovation*.

Applying a stochastic model component with the deterministic model can improve accuracy of the forecasts, provided the forecast errors of the deterministic model are autocorrelated.

Note 8.2:   The general principles of time series analysis will not be discussed here. The works of Box and Jenkins (1994) and Anderson (1976) are a great source for details on linear time series models (AR, MA, ARMAX), which are of importance to the present purpose. Neither will the iterative process of choosing the right model class be discussed here. Instead, forecast errors will be modeled by a simple AR process. Here it should be mentioned that other approaches, such as Bayesian learning algorithms, can also be applied for recursive predictions when using pure stochastic hydrologic models (Wood and Szöllősi-Nagy, 1981).

Let's assume that the prediction error sequence can be described by an $\text{AR}(\mu)$ model

$$\varepsilon_t = a_1\varepsilon_{t-\Delta t} + a_2\varepsilon_{t-2\Delta t} + \cdots + a_\mu\varepsilon_{t-\mu\Delta t} + w_{t-\Delta t} \tag{8.4}$$

where μ is the order (or memory) of the autoregressive process; $a_1, \ldots, a_\mu$ are its parameters, and w is a GWN sequence. Eq. 8.4 can be formulated in a state–space framework through the following definitions

$$x_{n+1,t} \triangleq \varepsilon_t \tag{8.5}$$
$$x_{n+2,t} \triangleq \varepsilon_{t-\Delta t}$$
$$\vdots$$
$$x_{n+\mu,t} \triangleq \varepsilon_{t-\mu\Delta t}.$$

The AR(μ) model now can be written as

$$
\begin{bmatrix} x_{n+1,t} \\ x_{n+2,t} \\ \vdots \\ x_{n+\mu,t} \end{bmatrix} = \begin{bmatrix} a_1 & a_2 & \cdots & a_\mu \\ 1 & 0 & \cdots & 0 \\ \vdots & \ddots & \ddots & \vdots \\ 0 & \cdots & 1 & 0 \end{bmatrix} \begin{bmatrix} x_{n+1,t-\Delta t} \\ x_{n+2,t-\Delta t} \\ \vdots \\ x_{n+\mu,t-\Delta t} \end{bmatrix} + \begin{bmatrix} 1 \\ 0 \\ \vdots \\ 0 \end{bmatrix} w_{t-\Delta t} \qquad (8.6)
$$

or equivalently

$$
\mathbf{x}'_t = \mathbf{\Phi}_\varepsilon(\Delta t)\mathbf{x}'_{t-\Delta t} + \mathbf{\Lambda}_\varepsilon w_{t-\Delta t}. \qquad (8.7)
$$

When measurement errors are assumed to be zero, an optimal estimate of the autoregressive parameters can be obtained (Szilágyi, 2004b) from the Yule-Walker equation

$$
\begin{bmatrix} 1 & r_{\varepsilon\varepsilon}(1) & \cdots & r_{\varepsilon\varepsilon}(\mu-1) \\ r_{\varepsilon\varepsilon}(1) & 1 & \ddots & \vdots \\ \vdots & \ddots & \ddots & r_{\varepsilon\varepsilon}(1) \\ r_{\varepsilon\varepsilon}(\mu-1) & \cdots & r_{\varepsilon\varepsilon}(1) & 1 \end{bmatrix} \begin{bmatrix} a_1 \\ a_2 \\ \vdots \\ a_\mu \end{bmatrix} = \begin{bmatrix} r_{\varepsilon\varepsilon}(1) \\ r_{\varepsilon\varepsilon}(2) \\ \vdots \\ r_{\varepsilon\varepsilon}(\mu) \end{bmatrix} \qquad (8.8)
$$

by inverting it

$$
\mathbf{a} = \mathbf{R}_\varepsilon^{-1}\mathbf{r}_\varepsilon \qquad (8.9)
$$

where $\mathbf{R}_\varepsilon$ is the correlation matrix of the prediction error sequence.

Note 8.3:   For an AR(1) sequence Eq. 8.9 yields

$$
a_1 = r_{\varepsilon\varepsilon}(1)
$$

while for an AR(2) it is

$$
a_1 = \frac{r_{\varepsilon\varepsilon}(1)[1 - r_{\varepsilon\varepsilon}(2)]}{1 - r_{\varepsilon\varepsilon}^2(1)}
$$

$$
a_2 = \frac{r_{\varepsilon\varepsilon}(2) - r_{\varepsilon\varepsilon}^2(1)}{1 - r_{\varepsilon\varepsilon}^2(1)}.
$$

For larger model-orders it is practical to use a numerical scheme.

Szilágyi (2004b) pointed out that the autoregressive parameter estimation above is correct only when no measurement error is present, which is never the case in practice. The presence of measurement error corrupts the autoregressive parameter estimation values obtained by the Yule-Walker equation. As a consequence, optimal estimates of the autoregressive parameters and thus optimal forecasts can only be obtained by the application of the Kalman filter during parameter estimation, which can be e.g. some systematic trial and error process. This way when measurement error is considered, the Yule-Walker equation has only limited practical value.

By augmenting the state vector of DLCM with the state vector of the prediction error sequence, Eq. 8.6, the *state vector of the deterministic–stochastic model* results as

$$\mathbf{x}_t^* = [x_{1,t}, \ldots, x_{n,t}, x_{n+1,t}, \ldots, x_{n+\mu,t}]^T = [\mathbf{x}_t, \mathbf{x}_t']^T. \tag{8.10}$$

With this, the combined, deterministic–stochastic state equation becomes

$$\mathbf{x}_t^* = \mathbf{\Phi}^*(\Delta t)\mathbf{x}_{t-\Delta t}^* + \mathbf{\Gamma}^*(\Delta t)u_{t-\Delta t} + \mathbf{\Lambda}^* w_{t-\Delta t} \tag{8.11}$$

where

$$\mathbf{\Phi}^*(\Delta t) = \begin{bmatrix} \mathbf{\Phi}(\Delta t) & \mathbf{0} \\ \mathbf{0} & \mathbf{\Phi}_\varepsilon(\Delta t) \end{bmatrix} \tag{8.12}$$

and

$$\mathbf{\Gamma}^*(\Delta t) = \begin{bmatrix} \mathbf{\Gamma}(\Delta t) \\ \mathbf{0} \end{bmatrix} \tag{8.13}$$

while

$$\mathbf{\Lambda}^* = [\underbrace{\mathbf{0}, 1, 0, \ldots, 0}_{\mu}]^T. \tag{8.14}$$

The output equation of the combined model can be written as

$$y_t = \mathbf{H}^* \mathbf{x}_t^* \tag{8.15}$$

where

$$\mathbf{H}^* = [\underbrace{0, 0, \ldots, k}_{n}, \underbrace{1, 0, \ldots, 0}_{\mu}]. \tag{8.16}$$

Eqs. 8.10 through 8.16 comprise one possibility of the *deterministic–stochastic model of streamflow forecasting in the pulse-data framework.* Similar equations can be written in the LI-data framework with the inclusion of stream–aquifer interactions by substituting the corresponding state and input-transition matrices together with the $\mathbf{\Omega}$ matrix.

8.2 RECURSIVE PREDICTION AND UPDATING

According to the definition of conditional predictions given in the Introduction section, the aim of forecasting is not only to give an estimate of a future streamflow value but also to specify the uncertainty of the estimate as well, because these two pieces of information together can help decision-makers with their risk analysis of alternative decisions. The task of forecasting this way becomes the estimation of future values of the state variables and specification of the expected forecast error variances in a way that they can be updated with the acquisition of the latest measurements.

Note 8.4: Such problems first occurred in control engineering in the early sixties in relation to spacecraft guidance. Rudolph Kalman worked out his well-known Kalman filter in 1960 for exactly these types of problems. The Kalman filter is a temporal generalization of the Wiener filter in a state–space framework description of stochastic systems. In essence, it gives a recursion for parameter estimation of conditional probability density functions. The first hydrological applications did not lag long behind (Hino, 1974; Szöllősi-Nagy, 1974) and quickly the Kalman filter found its way not only into hydrology but into hydraulics as well. However, many times it has been used as a fad, and often its potentials were overestimated. The Kalman filter is nothing more than a recursive algorithm, which facilitates optimal estimation and forecasting of measurable or directly nonmeasurable state variables of linear dynamic systems corrupted with additional noise. The emphasis is on optimality: it can be proved (Aoki, 1967; Meditch, 1969) that no other estimation algorithm can improve upon it when linear systems are concerned.

Eqs. 8.11 and 8.15 are in a form to solve the problem of conditional predictions and updating using the recursive Kalman filter algorithm. Derivation of the algorithm with the assumptions employed are described in Appendix I. The Kalman filter of Eqs. 8.11 and 8.15 is comprised of two alternately repeating steps:

(1) Δt lead-time forecasting of the state vector and the associated error covariance

$$\hat{\mathbf{x}}^*_{t|t-\Delta t} = \mathbf{\Phi}^*(\Delta t)\hat{\mathbf{x}}^*_{t-\Delta t|t-\Delta t} + \mathbf{\Gamma}^*(\Delta t)u_{t-\Delta t} \tag{8.17}$$

$$\mathbf{P}^*_{t|t-\Delta t} = \mathbf{\Phi}^*(\Delta t)\mathbf{P}^*_{t-\Delta t|t-\Delta t}\mathbf{\Phi}^{*T}(\Delta t) + \mathbf{\Lambda}^*Q_{t-\Delta t}\mathbf{\Lambda}^{*T} \tag{8.18}$$

(2) state variable and error covariance updating with the help of new measurement, z_t

$$\mathbf{K}^*_t = \mathbf{P}^*_{t|t-\Delta t}\mathbf{H}^{*T}[\mathbf{H}^*\mathbf{P}^*_{t|t-\Delta t}\mathbf{H}^{*T} + R_t]^{-1} \tag{8.19}$$

$$\hat{\mathbf{x}}^*_{t|t} = \hat{\mathbf{x}}^*_{t|t-\Delta t} + \mathbf{K}^*_t[z_t - \mathbf{H}^*\hat{\mathbf{x}}^*_{t|t-\Delta t}] \tag{8.20}$$

$$\mathbf{P}^*_{t|t} = [\mathbf{I}_{n+\mu} - \mathbf{K}^*_t\mathbf{H}^*]\mathbf{P}^*_{t|t-\Delta t} \tag{8.21}$$

where Q_t is the possibly time-dependent variance of w. $\mathbf{P}^*$ is the *a priori* or *a posteriori* covariance of the augmented state estimation error

$$\mathbf{P}^*_{.|.} = \begin{bmatrix} 0 & 0 \\ 0 & \mathbf{P}_{\varepsilon.|.} \end{bmatrix} \tag{8.22}$$

where $\mathbf{P}_{\varepsilon.|.}$ is a $\mu \times \mu$ covariance matrix of $\mathbf{x}'_{.|.}$, while R_t is the possibly time-dependent measurement error variance, a scalar. $\mathbf{I}_{n+\mu}$ here is an $(n + \mu) \times (n + \mu)$ identity matrix, and $\mathbf{K}^*_t$ is the $(n + \mu) \times (n + \mu)$ Kalman-gain matrix. The output equation is simply

$$\hat{\mathbf{y}}_{t|t-\Delta t} = \mathbf{H}^* \hat{\mathbf{x}}^*_{t|t-\Delta t} \tag{8.23}$$

while the variance of prediction error is

$$P^y_{t|t-\Delta t} = \mathbf{H}^* \mathbf{P}^*_{t|t-\Delta t} \mathbf{H}^{*T}. \tag{8.24}$$

A magnitude estimate can be obtained for $Q_t = Q$ (now a constant) by rearranging Eq. 8.4 as

$$\hat{w}_{t-\Delta t} = \varepsilon_t - \sum_{j=1}^{\mu} a_j \varepsilon_{t-j\Delta t} \quad t = i\Delta t, \quad i = \mu, \mu + 1, \ldots \tag{8.25}$$

and calculating the sample variance of $\hat{w}$. As was mentioned above, this estimate is inaccurate but may help to provide an initial estimate of Q for subsequent optimization of its value.

Note that in the augmented state variable case now, model uncertainty does not affect the original state variable, $\mathbf{x}_t$. This is so because the augmented state equation is made up of two separate submodels, a deterministic discrete cascade model, and an AR model that deals with model uncertainty, while additionally the Kalman filter takes care of the measurement error. Still, the advantage of applying an augmented state equation approach sofar is that model parameters, deterministic and stochastic alike, seem this way more naturally optimized together with the Kalman filter running during the optimization, which indeed this way results in optimized parameter estimates. On the other hand, if the deterministic model was to run separately, then it would be tempting to optimize the parameters of the cascade model first and subsequently optimize the AR model with or without (e.g. using the Yule-Walker equation) the Kalman filter running. As was pointed out by Szilágyi (2004b), it is imperative to optimize *all model parameters with the Kalman filter running during optimization* in order to truly obtain optimal model parameter values.

Multi-step predictions can be achieved by inserting the *a priori* one-step prediction of the augmented state variable into Eq. 5.77 or 5.80 if input forecasts are available (similarly into Eq. 6.21 or 7.20)

$$\hat{y}_{t+i\Delta t|t-\Delta t} = \mathbf{H}^*\mathbf{\Phi}^{*i}(\Delta t)\hat{\mathbf{x}}^*_{t|t-\Delta t}$$

$$+ \left[\mathbf{H}^* \sum_{j=0}^{i-1} \mathbf{\Phi}^{*j}(\Delta t)\mathbf{\Gamma}^*(\Delta t) \right] u_{t-\Delta t} \quad i = 1, \ldots \quad (8.26)$$

with the corresponding variance of prediction error as

$$P^y_{t+i\Delta t|t-\Delta t} = \mathbf{H}^*\mathbf{P}^*_{t+i\Delta t|t-\Delta t}\mathbf{H}^{*T} \tag{8.27}$$

where (Meditch, 1969)

$$\mathbf{P}^*_{t+i\Delta t|t-\Delta t} = \mathbf{\Phi}^{*i}(\Delta t)\mathbf{P}^*_{t|t-\Delta t}(\mathbf{\Phi}^{*T})^i(\Delta t)$$

$$+ \sum_{j=0}^{i-1} \mathbf{\Phi}^{*j}(\Delta t)\mathbf{\Lambda}^*Q_{t-\Delta t}\mathbf{\Lambda}^{*T}(\mathbf{\Phi}^{*T})^j \quad i = 1, \ldots . \tag{8.28}$$

The Kalman filter algorithm requires estimates for the following terms: Q_t, R_t, $\hat{\mathbf{x}}^*_{0|0}$, and $\mathbf{P}^*_{0|0}$. From these four terms, specifying Q_t and R_t accurately is the most important because these values are not updated by the filter. If Q_t is assumed to be constant in time, then the w estimates of Eq. 8.25 can help with the Q term's initialization. $\hat{\mathbf{x}}^*_{0|0}$ can be constructed by the initial value, $\mathbf{x}_0$, obtained from Eq. 5.69 (or Eq. 6.20 with or without the $\mathbf{\Omega}$ term, respectively) plus by an initial guess of the AR parameters. The $\mathbf{P}^*_{0|0}$ term can be initialized with a sample covariance matrix of model errors.

As evidenced by Eqs. 8.19 through 8.21, the predictions are updated recursively with the arrival of new observations in each sampling instant. The *a posteriori* estimate of state is achieved through a linear weighting of the *a priori* state estimate and the new observation (see Eq. 8.20). It is important to have a measurement variance different from zero for this weighting to work. When R_t is assumed to be constant and zero, the *a posteriori* state estimation becomes equal to $\mathbf{H}^{*-1}z_t$ (Ahsan and O'Connor, 1994; Szilágyi, 2004b), and in such a case application of the Kalman filter during parameter estimation reduces to a traditional time series parameter estimation, yielding the same estimates as the Yule-Walker equation for an autoregressive process (see Appendix I). However, since measurement uncertainty is always present with a variance larger than zero, the application of the Kalman filter during optimization is always expected to result in better parameter estimates and so in more accurate predictions than traditional parameter estimation techniques (Szilágyi, 2004b).

Note 8.5:   Because the DLCM is a SISO (single input/single output) system, the term to be inverted in Eq. 8.19 is just a scalar.

Note 8.6:   The Kalman filter is not used for real-time updating of any of the, deterministic or stochastic, model parameters. Model parameters, instead, are optimized off-line and even then not in a parameter updating mode. When performing off-line optimization, a set of values is prescribed systematically for the model parameters and kept constant over the optimization period. With each set of parameter values, a mean-square error is calculated for the optimization period before a new set of values is prescribed for the parameters. The optimization stops when the parameter values have spanned the prescribed parameter space with a predefined resolution. The parameter values that belong to the smallest mean-square-error are retained and considered to be optimal. As was pointed out above, even this off-line optimization should be carried out with the Kalman filter running during the optimization process in order to obtain fully optimal model parameters.

In contrast, coupled, real-time **parameter** *and* **state** updating is a nonlinear optimization process (Eykhoff, 1974) and its linearization (as is the Extended Kalman Filter [EKF]) brings with it certain unwanted properties such as noise sensitivity and possible divergence. Therefore application of the EKF will not be discussed here.

It could be argued that there was no need to formulate the deterministic model component in a state–space framework if in the end the Kalman filter is applied over an additional and, in fact, separate stochastic model component only (see Eq. 8.22). Indeed, the objective of writing the deterministic model in a state–space form was motivated by the goal of applying the Kalman filter over the deterministic model itself. If the autocorrelation of prediction errors is insignificant, the Kalman filter can be straightforwardly applied with the deterministic model as described in Appendix I. Fortunately, the same can be achieved even when forecast or model errors are correlated, without needing to apply a separate stochastic AR model component demonstrated above.

The solution again requires state augmentation. The state and measurement equations (see Eq. A2.8) can now be written as

$$\mathbf{x}_t = \mathbf{\Phi}(\Delta t)\mathbf{x}_{t-\Delta t} + \mathbf{\Gamma}(\Delta t)u_{t-\Delta t} + \mathbf{\Gamma}_\nu \boldsymbol{v}_{t-\Delta t} \qquad (8.29)$$

$$\boldsymbol{v}_t = \boldsymbol{\varphi}\boldsymbol{v}_{t-\Delta t} + \mathbf{w}_t^{(1)} \qquad (8.30)$$

$$z_t = \mathbf{H}\mathbf{x}_t + w_t^{(2)} \qquad (8.31)$$

where $\boldsymbol{v}$ $(n \times 1)$ is assumed to be a normally distributed, vector-valued, first-order autoregressive [AR(1)] sequence (also called a Gauss-Markov sequence) of model errors (Meditch, 1969; Bras and Rodriguez-Iturbe, 1993). $\boldsymbol{\varphi}$ $(n \times n)$ is the diagonal matrix of the AR(1) parameters, z is the measured output, and $\mathbf{w}^{(1)}$ $(n \times 1)$ and $w^{(2)}$ are GWN sequences, the latter is called the *measurement error*. The model error distribution matrix, $\mathbf{\Gamma}_\nu$ $(n \times n)$, is now an identity matrix.

The augmented variables and system matrices can be written as

$$\mathbf{x}_t^* = \begin{bmatrix} \mathbf{x}_t \\ \boldsymbol{v}_t \end{bmatrix}; \quad \boldsymbol{\Phi}^*(\Delta t) = \begin{bmatrix} \boldsymbol{\Phi}(\Delta t) & \boldsymbol{\Gamma}_v \\ 0 & \varphi \end{bmatrix}; \quad \boldsymbol{\Lambda}^* = \begin{bmatrix} 0 \\ \mathbf{I} \end{bmatrix}; \quad (8.32)$$

$$\boldsymbol{\Gamma}^*(\Delta t) = \begin{bmatrix} \boldsymbol{\Gamma}(\Delta t) \\ 0 \end{bmatrix}; \quad \mathbf{H}^* = \begin{bmatrix} \mathbf{H} & 0 \end{bmatrix}.$$

where $\mathbf{I}$ is another $(n \times n)$ identity matrix. The dimensions of the augmented variables (from left to right by row) are: $2n \times 1$, $2n \times 2n$, $2n \times n$, $2n \times 1$, and $1 \times 2n$, respectively. Hence, the augmented state and measurement equations become

$$\mathbf{x}_t^* = \boldsymbol{\Phi}^*(\Delta t)\mathbf{x}_{t-\Delta t}^* + \boldsymbol{\Gamma}^*(\Delta t)u_{t-\Delta t} + \boldsymbol{\Lambda}^*\mathbf{w}_t^{(1)} \tag{8.33}$$

$$z_t = \mathbf{H}^*\mathbf{x}_t^* + w_t^{(2)}. \tag{8.34}$$

Eqs. 8.17 through 8.21 again can be used for conditional one-step forecasting and updating. The $\mathbf{Q}_t$ term of Eq. 8.18 now becomes a (time-dependent) covariance matrix of the noise term, $\mathbf{w}^{(1)}$. R_t now is the (time-dependent) variance of $w^{(2)}$, while $\mathbf{P}_t^*$ is the $(2n \times 2n)$ *a priori* or *a posteriori* covariance matrix of the augmented state variable

$$\mathbf{P}_t^* = \begin{bmatrix} \mathbf{P_{xx}} & \mathbf{P_{xv}} \\ \mathbf{P}_{xv}^T & \mathbf{P}_{vv} \end{bmatrix} \tag{8.35}$$

where all covariances within the $\mathbf{P}_t^*$ matrix are time-dependent.

The filter algorithm again requires the specification of the terms in $\mathbf{Q}_t$, $\sigma_{w^{(2)}}^2$ for R_t, as well as the $\mathbf{P_{xx}}$, $\mathbf{P_{xv}}$, and $\mathbf{P}_{vv}$ terms for $\mathbf{P}_0^*$. An initial value of $\mathbf{P_{xx}}$ may be estimated as

$$\sigma_{x_i x_j}^2 = K^2 \sigma_y^2 \quad i,j = 1,\ldots,n \tag{8.36}$$

where K is the mean storage delay time, $K = k^{-1}$ of the storage element. Each $\mathbf{Q}_t$ term (plus the the diagonal terms of $\mathbf{P}_{vv}$ for $t = 0$) was estimated as $(0.04\,u_t)^2$, while $\sigma_{w^{(2)}}^2$ as 10% of the former. The initial value of $\mathbf{P_{xv}}$ was set to zero, as well as the off-diagonal terms of $\mathbf{P}_{vv}$. Through trial and error the value of $\varphi_1 = \varphi_2$ became 0.7 for data in Fig. 8.1.

Figs. 8.1 through 8.6 demonstrate the effect of the Kalman filter on the one-step (24-hour) forecasts using the stations of Figs. 7.2 and 7.3. The time-period is now a subset of that of Fig. 7.2. As it can be seen, the deterministic model prediction errors are highly correlated. The Kalman filter, using the augmented state approach of Eqs. 8.32 through 8.34, greatly reduces this autocorrelation, making the filtered forecast errors

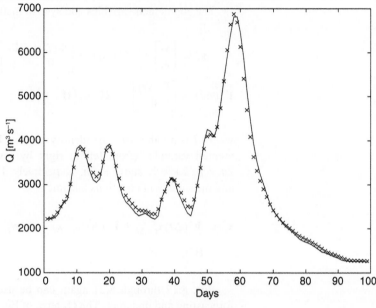

Figure 8.1. A subset of Fig. 7.3 for 1-day forecasts.

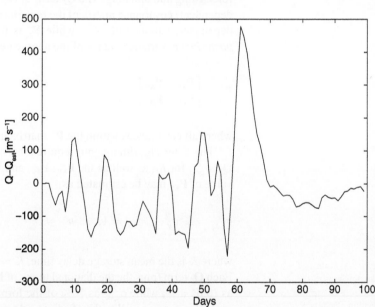

Figure 8.2. Error sequence of the 1-day forecasts.

become a GWN sequence. The mean standard deviation of the deterministic model forecast error of 126 $m^3 s^{-1}$ was reduced to 53 $m^3 s^{-1}$ through the application of the filter. Fig. 8.4 also displays the standard deviation of prediction error for each individual forecast. Since both model and measurement errors are assumed to be directly proportional to the input, these intervals widen with increasing flow values. Note the initially large forecast uncertainty as a result of inaccurate estimation of $\mathbf{P}_0^*$.

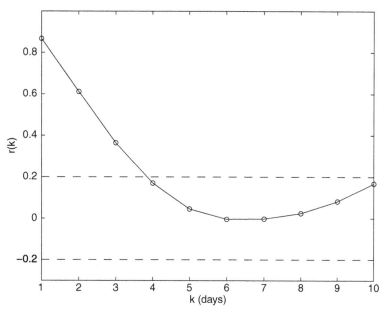

Figure 8.3. Autocorrelation (r) function of the 1-day forecast errors. Also displayed is the 95% confidence interval for $r = 0$.

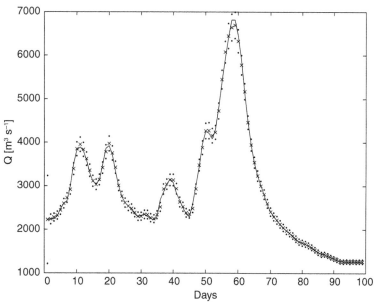

Figure 8.4. Kalman filtered 1-day forecasts of Fig. 8.1 with the corresponding standard deviation of errors.

As before, multi-step predictions with the corresponding variances of prediction error can be obtained from Eqs. 8.26 through 8.28.

In the LI-data system framework with or without stream–aquifer interactions, the above filter-steps (Eqs. 8.17 through 8.21 and 8.26 through 8.28) remain valid after inclusion of the corresponding extra terms in the state equation, as was done in the example.

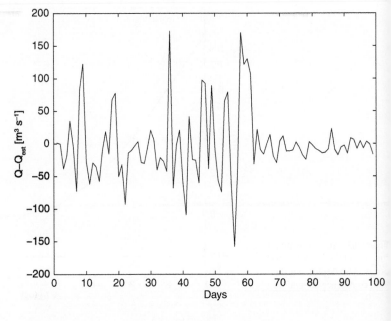

Figure 8.5. Error sequence of the Kalman filtered 1-day forecasts.

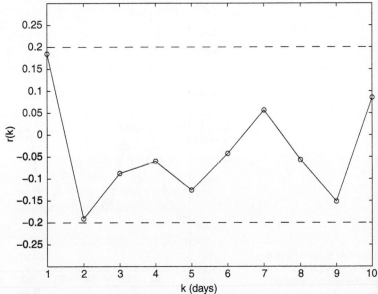

Figure 8.6. Autocorrelation (r) function of the 1-day forecast errors. Also displayed is the 95% confidence interval for $r = 0$.

In this chapter the stochastic component of DLCM was discussed. The error sequence is described by an AR process and is formulated in a state–space framework which enables the construction of an augmented deterministic–stochastic model. Recursive prediction and updating of the augmented state is performed by the linear Kalman filter through a continuous feedback of the prediction error. Conditional prediction of the output is achieved by a linear projection of the a priori *augmented state variable.*

EXERCISES

8.1. Try out the Kalman filter algorithm on a scalar AR(1) process you generate with the computer. Let the model and measurement errors be Gaussian white noises. Estimate the AR(1) parameter first with the Yule-Walker equation, then with systematic trial and error while the Kalman filter is running. Which parameter estimate yields better result? What happens when there is no measurement error? Which method gives better forecasts?

CHAPTER 9

Some Practical Aspects of Model Application for Real-Time Operational Forecasting

Below some of the practical considerations about model parameters, their optimization and sensitivity are discussed. The coupled deterministic–stochastic model is compared to a pure stochastic approach in terms of model accuracy and practicality. Finally, a concrete example is given on how the model is set up for operational real-time forecasting of flow rates and water stages of the Danube and its major tributaries in Hungary.

9.1 MODEL PARAMETERIZATION

Optimization of the model parameters (n, k, g, and C_0) can be achieved by numerous techniques (see e.g. Press et al., 1986). Harkányi (1982) worked out a special algorithm for the optimization of the DLCM parameters without stream–aquifer interactions. His direct technique does not require derivatives and uses the ordinary least-squares expression as the target function (J) to be minimized

$$J = \sum_t (\widehat{y}_t - y_y)^2 \longrightarrow \min_{(n,k)}. \tag{9.1}$$

The resulting parameters will be valid for both, low- and high-flow periods.

Note 9.1: During floods, the value of the storage coefficient ($K = k^{-1}$) may change significantly due to a marked difference in the friction coefficient's value between the main channel and the flood-plain. For such problems Becker and Glos (1970) worked out their Critical Level Model (CLM), where the flood discharge can be divided into different discharge intervals and the resulting discharges separately routed through their corresponding linear submodels, all connected in parallel. Ambrus et al. (1984) report of a study where the DLCM was incorporated into a CLM for a tributary of the Danube.

With modern personal computers, recalculation of the forecasts for a given period, using different trial-values of the parameters, can be done extremely fast within the recursive state–space approach. Generally, it takes only seconds, to systematically try out all possible combinations of the model parameters once an interval and a corresponding increment is defined for each parameter. The optimization starts with a predefined minimum value of each parameter which is systematically incremented until an arbitrary maximum value is reached for all parameters and, correspondingly, all possible variations of the parameter values have been exhausted with the chosen resolution. The combination of the parameter values that minimizes Eq. 9.1 is considered as the optimal set of the parameters. The result of such direct trial and error optimization, although probably the most time consuming of all available optimization techniques, depends only on the assigned resolution (i.e. increment) of each parameter but gives a true optimum that is no longer a function of the chosen optimization method. When the parameters have physical meaning, as with DLCM, assigning a possible lower and upper limit for each parameter value is self-evident. The prescribed resolution can be a sole function of computer power.

Experiments conducted at the National Hydrological Forecasting Service of Hungary (NHFSH) indicated that the n and k parameters of the model are remarkably stable, their recursive updating is not necessary. The model is more sensitive to the n value than to the value of k, which can partially be explained by the fact that the former parameter can traditionally take only integer values, although the model's structure could allow for non-integer n values. A noninteger n version of DLCM (Szilágyi, 2006) has been discussed in detail previously. Thus a change from $n = 1$ to $n = 2$ immediately means a 100% increment in parameter value. Similar experiments with the g and C_0 parameters have yet to be accomplished.

9.2 COMPARISON OF A PURE STOCHASTIC, A DETERMINISTIC (DLCM), AND DETERMINISTIC–STOCHASTIC MODELS

The problem of flow forecasting can be tackled by using a "black box" approach, where the physics behind the stream-flow process is not defined explicitly. Similarly to Chapter 8.1, a pure, stochastic ARMA model may assume the following linear relationship between in- (u) and outflow (y) values of a stream reach

$$y_t = a_{1,t} y_{t-1} + a_{2,t} y_{t-2} + \cdots + a_{n,t} y_{t-n} + b_{1,t} u_{t-1}$$
$$+ b_{2,t} u_{t-2} + \cdots + b_{m,t} u_{t-m} + v_t \tag{9.2}$$

where m and n are the number of past in- and outflow values that affect the outflow at time t; v is a GWN sequence with zero mean and given variance; while a_i and b_i are the unknown time-dependent ARMA coefficients.

By defining the following vector variables

$$\mathbf{\Theta}_t = [a_{1,t}, a_{2,t}, \ldots, a_{n,t}, b_{1,t}, b_{2,t}, \ldots, b_{m,t}]^T \tag{9.3}$$

and

$$\mathbf{H}_t = [y_{t-1}, y_{t-2}, \ldots, y_{t-n}, u_{t-1}, u_{t-2}, \ldots, u_{t-m}] \tag{9.4}$$

the above equation can be written as

$$y_t = \mathbf{H}_t \mathbf{\Theta}_t + v_t \tag{9.5}$$

which describes the output equation of a time-variant, discrete dynamic system with state variable $\mathbf{\Theta}$. Since the value of the state variable changes through time in an *a priori* unknown fashion, Szöllősi-Nagy et al. (1977) assumed this change to be a Gauss-Markov sequence

$$\mathbf{\Theta}_t = \mathbf{\Theta}_{t-1} + \mathbf{w}_t \tag{9.6}$$

where $\mathbf{w}$ is again a GWN sequence. Note that Eq. 9.6 is Eq. A2.1 with $\mathbf{\Phi}_t = \mathbf{I}$ and $\mathbf{\Gamma}_t = \mathbf{0}$. The estimation of the state variable, $\mathbf{\Theta}_t$, can be achieved with the help of the Kalman filter. In order to avoid a nonlinear estimation of both, the state variable and the noise statistics, $\mathbf{Q}_t$ and R_t, the latter statistics can be estimated off-line with a trial and error approach over a suitably long period and taken to be constant in time (Szöllősi-Nagy and Mekis, 1982). The other possibility is to use a nonlinear estimation approach described by Young (1984).

The one-step forecast of outflow is obtained by taking the expectation of Eq. 9.6

$$\bar{y}_{t|t-1} = \mathbf{H}_t \overline{\mathbf{\Theta}}_{t-1|t-1} \tag{9.7}$$

regardless of the method by which the *a priori* estimate of the state variable is obtained. These forecasts, obtained by the linear estimation approach, were compared with forecasts of the DLCM and its coupled, deterministic–stochastic model version at NHFSH.

The ARMA model with its optimized model-order of $N = n + m = 8$ performed the worst of the three models, while the coupled, deterministic–stochastic model the best. Another disadvantage presented by the pure ARMA model, beside its poorer performance, is that it requires significantly more parameters than the deterministic–stochastic model. Note that even the extended DLCM with its four parameters to account for stream–aquifer interactions, has half the number of parameters than the above ARMA model. And this is only for the one-step forecast, because for each lead-time, the ARMA model has to be re-parameterized (Young, 2002); thus, for a typical 1–4-day forecast scenario it immediately means

32 parameters to be optimized, as opposed to the constant number of four parameters for the deterministic–stochastic model.

For an illustration see Tables 9.1 and 9.2, where several statistics of the measured stream flows and their one-day forecasts for Dunaföldvár (Fig. 5.11) are displayed.

Here the DLCM was run in a pulse-data framework and no stream–aquifer interactions were accounted for, i.e. $g = C_0 = 0$. $\bar{\varepsilon}(1)$ is the average difference between observed and forecasted flow values (i.e. forecast error) with a lead-time of one day, and the corresponding standard deviation is $\sigma_\varepsilon(1)$. $r_\varepsilon(1)$ is the autocorrelation value of the one-day forecast error. Finally, the efficiency coefficient, $\eta_\varepsilon(k)$, is defined as

$$\eta_\varepsilon(k) = \sqrt{1 - \left(\frac{\sigma_\varepsilon(k)}{\sigma_\Delta(k)}\right)^2} \qquad (9.8)$$

where $\sigma_\Delta(k)$ is the standard deviation of the change in the measured flow values

$$\Delta_t(k) = y_t - y_{t+k} \qquad (9.9)$$

during the forecast period.

Rainfall as an MA-process in state–space

Example 9.1: To model rainfall sequences a moving average (MA) model of order n

$$y(t) = \Theta_1 w(t-1) + \Theta_2 w(t-2) + \cdots + \Theta_n w(t-n)$$

is frequently used in hydrology (e.g. Matalas, 1963), where the Θs are the moving-average parameters and $w(\cdot)$ is the GWN sequence. Defining

Table 9.1. One-day forecast $[m^3 s^{-1}]$ statistics for Dunaföldvár (1980) by different models.

Statistics:	$\bar{\varepsilon}(1)$	$\sigma_\varepsilon(1)$	$r_\varepsilon(1)$	$\eta_\varepsilon(1)$
ARMA	0.78	200.0	−0.03	0.61
DLCM	−111.3	110.6	0.74	0.71
DLCM + stochastic	−5.69	78.8	0.08	0.85

Table 9.2. Mean $(\bar{y})$, standard deviation (σ_y) and one-step autocorrelation coefficient $[r(1)]$ of the measured daily instantaneous flow values $[m^3 s^{-1}]$ at Dunaföldvár (1980) and their one-day forecasts.

Statistics:	$\bar{y}$	σ_y	$r(1)$
Measured	2352	846	0.98
ARMA	2343	889	0.94
DLCM	2463	803	0.97
DLCM + stochastic	2358	861	0.97

the state variables as $x_1(t) = w(t-n)$, $x_2(t) = w(t-n+1)$, ... , $x_n(t) = w(t-1)$, the above equation can be written as

$$\mathbf{x}(t+1) = \boldsymbol{\Phi}x(t) + \boldsymbol{\Gamma}w(t)$$

where

$$\boldsymbol{\Phi} = \begin{bmatrix} 0 & 1 & 0 & \cdots & 0 \\ 0 & 0 & 1 & \cdots & 0 \\ \vdots & & & \ddots & \vdots \\ 0 & 0 & & 0 & 1 \\ 0 & 0 & \cdots & 0 & 0 \end{bmatrix}, \quad \boldsymbol{\Gamma} = \begin{bmatrix} 0 \\ 0 \\ \vdots \\ 1 \end{bmatrix}$$

and

$$y(t) = \mathbf{H}x(t)$$

with

$$\mathbf{H} = [\Theta_n, \Theta_{n-1}, \ldots, \Theta_1].$$

An ARMA-process
in state–space

Example 9.2: Here an alternate state–space representation of an ARMA model is given. Consider the ARMA(n, m) model

$$y_{t+1} + \Phi_1 y_t + \Phi_2 y_{t-1} + \cdots + \Phi_n y_{t-n+1}$$
$$= \Theta_1 w_t + \Theta_2 w_{t-1} + \cdots + \Theta_m w_{t-m+1}$$

with $\mathbf{x}_t = [x_{t-n+1}, x_{t-n+2}, \ldots, x_t]^T$ and $\mathbf{w}_t = [w_t, w_{t-1}, \ldots, w_{t-m+1}]^T$, so that the state–space model can be written as

$$\mathbf{x}_{t+1} = \boldsymbol{\Phi}\mathbf{x}_t + \boldsymbol{\Gamma}w_t$$
$$y_t = \mathbf{H}\mathbf{x}_t$$

with

$$\boldsymbol{\Phi} = \begin{bmatrix} 0 & 1 & & 0 \\ \vdots & & \ddots & \\ 0 & 0 & & 1 \\ -\Phi_n & -\Phi_{n-1} & \cdots & -\Phi_1 \end{bmatrix}$$

$$\boldsymbol{\Gamma} = \begin{bmatrix} 0 & 0 & \cdots & 0 \\ \vdots & & & \vdots \\ 0 & & & 0 \\ \Theta_1 & \Theta_2 & \cdots & \Theta_m \end{bmatrix}, \quad \mathbf{H} = \begin{bmatrix} 1 \\ 0 \\ \vdots \\ 0 \end{bmatrix}^T$$

where $\boldsymbol{\Gamma}$ is an $n \times m$ matrix.

9.3 APPLICATION OF THE DETERMINISTIC–STOCHASTIC MODEL FOR THE DANUBE BASIN IN HUNGARY

The coupled, deterministic–stochastic model started its operative service at NHFSH in 1983. Typically, it produces stage and flow forecasts on a daily basis, but during flood events, forecasts can be issued/updated at 12-h intervals. The model uses stage measurements taken at 6 a.m. each day. The stage measurements are converted into instantaneous flow rates using a rating curve for each gauging station. Forecasts, both in stage and flow-rate forms, are generally ready and distributed to the relevant agencies by 10 a.m. and can be looked up/downloaded from the Service's website.

Fig. 9.1 displays the logical structure of forecasts for the major gauging stations of the Danube in Hungary (Fig. 9.2), omitting tributaries.

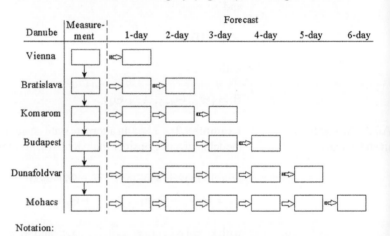

Figure 9.1. Forecast structure for the Danube in Hungary.

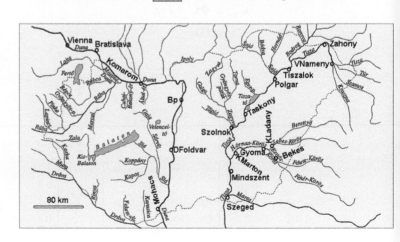

Figure 9.2. Stream network of Hungary.

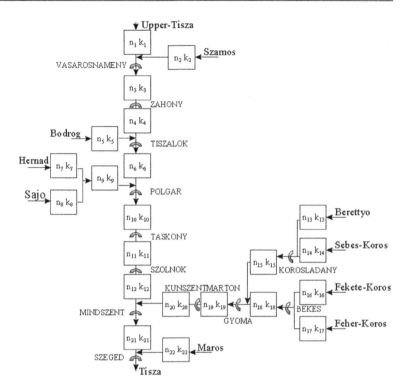

Figure 9.3. System of linear cascades for the Tisza River in Hungary.

Finally, the model structure is depicted for the largest tributary of the Danube, the Tisza River with its sub-tributaries, in Fig. 9.3. The names in capitals denote towns where the gauging stations are located.

Each cascade is represented by two parameters, n and k, provided $g = C_0 = 0$ for each cascade. Whether accounting for stream–aquifer interactions improves forecast accuracy and reliability, will be the focus of future investigations.

Harkányi and Bartha (1984) applied the DLCM for rainfall–runoff modeling. Nonlinearity of the process was accounted for by using an antecedent precipitation index (API) in the transformation. They showed that the runoff ratio and API is related through a gamma distribution. The model, $\Gamma(API)$, generates input to DLCMs connected in parallel to model surface and the sub-surface runoff. This way runoff is predicted from measured precipitation for the uppermost gauging stations of the Danube's tributaries.

The coupled, deterministic–stochastic model described in this study has been in operational use (outside Hungary) for several years in Thailand, Malaysia, and Germany.

Summary

This study focused on real-time forecasting of stream flow by a coupled, deterministic–stochastic model.

The first chapter defined the scope of the study and explained the reasons that called for such an approach. A probabilistic definition of forecasting has also been specified.

The second chapter gave a brief tally of the continuous flow routing techniques. It was pointed out that these linear models with constant wave speed are all obtainable through a discretization of the continuous linear kinematic wave. Continuity, steady state, and transitivity were defined in the following chapters. The properties of the continuous cascade are summarized below.

Thesis 1: The time-invariant dynamic system of the continuous KMN-cascade is defined by the

$$\dot{\mathbf{x}}(t) = \mathbf{F}\mathbf{x}(t) + \mathbf{G}u(t)$$

$$y(t) = \mathbf{H}\mathbf{x}(t)$$

state and output equations, where

$$[F]_{i,j} = \begin{cases} -k, & i = j \\ k, & i = j - 1; \quad i = 1, 2, \dots, n \\ 0, & \text{otherwise} \end{cases}$$

$$\mathbf{G} = [1, 0, \dots, 0]^T$$

$$\mathbf{H} = [0, 0, \dots, k]$$

with $k = K^{-1}$, where K is the mean delay time of the characteristic reach. The continuous cascade is unambiguously defined by the

$$\Sigma_{KMN} = (\mathbf{F}, \mathbf{G}, \mathbf{H})$$

matrix-triplet. The impulse response of the KMN-cascade thus becomes

$$h(t) = k(kt)^{n-1} \frac{1}{(n-1)!} e^{-kt}.$$

The continuous KMN-cascade is equivalent to the continuous, spatially discrete, linear kinematic wave. Continuity and transitivity unconditionally apply to the continuous KMN-cascade where storage is the same in each storage element in a steady state.

The discrete version of the continuous KMN-cascade was derived in both pulse-data and LI-data system frameworks. It was shown that a trivial discretization of the continuous KMN-cascade is not adequate. The conditionally adequate discrete model (DLCM) was obtained by integrating the state-trajectory over a predefined constant Δt sampling time-interval. It was shown that the discrete model is discretely coincident with its continuous counterpart, preserves unconditional continuity, and is transitive in the $\Delta t \longrightarrow 0$ limit. These results are summarized below.

Thesis 2: Within the pulse-data framework, the state and output equations of the $\Sigma_{DLCM}(\Delta t) = [\mathbf{\Phi}(\Delta t), \mathbf{\Gamma}(\Delta t), \mathbf{H}]$ discrete version of the Σ_{KMN} continuous cascade are

$$\mathbf{x}_{t+\Delta t} = \mathbf{\Phi}(\Delta t)\mathbf{x}_t + \mathbf{\Gamma}(\Delta t)u_t$$

$$y_t = \mathbf{H}\mathbf{x}_t$$

where

$$[\mathbf{\Phi}(\Delta t)]_{i,j} = \begin{cases} \dfrac{(k\Delta t)^{i-j}}{(i-j)!}e^{-k\Delta t}, & i \geq j \\ 0, & i < j \end{cases}$$

$$[\mathbf{\Gamma}(\Delta t)]_i = \frac{1}{k}\left(1 - e^{-k\Delta t}\sum_{j=0}^{i-1}\frac{(k\Delta t)^j}{j!}\right) = \frac{1}{k}\frac{\Gamma(i, k\Delta t)}{\Gamma(i)}$$

$$[H]_j = \begin{cases} 0, & j \neq n \\ k, & j = n \end{cases}.$$

Any two conditionally adequate discrete models of time-intervals Δt and $\Delta t^* = \mu\Delta t$ are linked by the following linear transformation

$$\Sigma_{DLCM}(\Delta t) \xrightarrow{\mathbf{T}_\Phi(\mu), \mathbf{T}_\Gamma(\mu)} \Sigma_{DLCM}(\Delta t^*)$$

where

$$[T_\Phi(\mu)]_{i,j} = \begin{cases} \dfrac{[(\mu - 1)k\Delta t]^{i-j}}{(i-j)!}e^{-k\Delta t(\mu-1)}, & i \geq j \\ 0, & i < j \end{cases}$$

$$< T_\Gamma(\mu) >_i = \frac{\Gamma(i, \mu k\Delta t)}{\Gamma(i, k\Delta t)}.$$

Within the LI-data framework, the state and output equations of the $\Sigma_{DLCM}(\Delta t) = [\mathbf{\Phi}(\Delta t), \mathbf{\Gamma}_1(\Delta t), \mathbf{\Gamma}_2(\Delta t), \mathbf{H}]$ discrete version of the Σ_{KMN}

continuous cascade are

$$\mathbf{x}_{t+\Delta t} = \mathbf{\Phi}(\Delta t)\mathbf{x}_t + \mathbf{\Gamma}_1(\Delta t)u_t + \mathbf{\Gamma}_2(\Delta t)u_{t+\Delta t}$$

$$y_t = \mathbf{H}\mathbf{x}_t$$

where $[\mathbf{\Phi}(\Delta t)]_{i,j}$ remains as above, and

$$[\mathbf{\Gamma}_1(\Delta t)]_i = \frac{1}{k}\frac{\Gamma(i, k\Delta t)}{\Gamma(i)}\left(\frac{i}{k\Delta t} - \frac{(k\Delta t)^{i-1}e^{-k\Delta t}}{\Gamma(i, k\Delta t)}\right)$$

$$[\mathbf{\Gamma}_2(\Delta t)]_i = \frac{1}{k}\frac{\Gamma(i, k\Delta t)}{\Gamma(i)}\left(1 + \frac{(k\Delta t)^{i-1}e^{-k\Delta t}}{\Gamma(i, k\Delta t)} - \frac{i}{k\Delta t}\right).$$

Any two conditionally adequate discrete models of time-intervals Δt and $\Delta t^* = \mu\Delta t$ are now linked by the following linear transformation

$$\mathbf{\Sigma}_{DLCM}(\Delta t) \xrightarrow{\mathbf{T}_\Phi(\mu), \mathbf{T}_{\Gamma_1}(\mu), \mathbf{T}_{\Gamma_2}(\mu)} \mathbf{\Sigma}_{DLCM}(\Delta t^*)$$

where $\mathbf{T}_\Phi(\mu)$ remains as above, and

$$< T_{\Gamma_1}(\mu) >_i = \frac{1}{\mu}\frac{i\Gamma(i, \mu k\Delta t) - (\mu k\Delta t)^i e^{-\mu k\Delta t}}{i\Gamma(i, k\Delta t) - (k\Delta t)^i e^{-k\Delta t}}$$

$$< T_{\Gamma_2}(\mu) >_i = \frac{1}{\mu}\frac{i\Gamma(i, \mu k\Delta t)(\mu k\Delta t - i) + (\mu k\Delta t)^i e^{-\mu k\Delta t}}{i\Gamma(i, k\Delta t)(k\Delta t - i) + (k\Delta t)^i e^{-k\Delta t}}.$$

The pulse-data system is a special case of the LI-data framework through the $u_{t+\Delta t} \triangleq u_t$ choice at time t.

In the pulse-data system framework the system-characteristic functions of DLCM are the unit-pulse and unit-step responses, while in the LI-data framework the unit-pulse response is replaced by two (one with a positive and one with a negative slope) unit-ramp response functions. This is so because any linear change from a to b over a predefined Δt interval can be described as the sum of two linear ramp functions: one that starts from unity at t and reaches zero Δt later, multiplied by a, and one that starts from zero at t and reaches unity over the same time-interval, and multiplied by b.

Thesis 3: The $\mathbf{\Sigma}_{DLCM}(\Delta t)$ conditionally adequate discrete cascade is observable, if the

$$[\Theta_n]_{i,j} = k\frac{(ik\Delta t)^{n-j}}{(n-j)!}e^{-ik\Delta t}$$

non-singular observation matrix has a rank equal to the order of the discrete cascade (n), provided $n \geq 1$, k and $\Delta t > 0$.

In practical applications, information of the initial state ($\mathbf{x}_0$) under non-permanent conditions is of importance.

Thesis 4:   The $\Sigma_{DLCM}(\Delta t)$ conditionally adequate discrete cascade's initial state can be unambiguously obtained from n pairs of inflow values and outflow values in the pulse-data system, and $n+1$ inflow values and n outflow values in the LI-data framework case as

$$\mathbf{x}_0 = \boldsymbol{\Theta}_n^{-1} \mathbf{e}_n$$

where

$$[e_n]_i = y_i - \sum_{j=0}^{i-1} h_{i-j} u_j$$

in the former and

$$[e_n]_i = y_i - \mathbf{H}\left[\sum_{j=1}^{i} \left(\boldsymbol{\Phi}^{i-j}(\Delta t)\boldsymbol{\Gamma}_1(\Delta t)u_{j-1} + \boldsymbol{\Phi}^{i-j}(\Delta t)\boldsymbol{\Gamma}_2(\Delta t)u_j \right) \right]$$

in the latter case. Here h_i is the *ith* ordinate of the unit-pulse response.

A recursive algorithm was given for the DCLM forecasts, with their asymptotic behavior specified. Another algorithm was derived for solving the inverse problem of forecasting: input detection.

Thesis 5:   The prediction-error sequence of the DLCM was modeled by a separate m-order autoregressive, AR(m), process, written in a state–space form; and, as an alternative, by the help of state-augmentation where the prediction-error sequence was considered as a Gauss-Markov process. Conditional prediction of the augmented state and its updating was performed by the linear Kalman filter algorithm. Conditional prediction of the flow was obtained by a linear projection of the *a priori* augmented state variable. By repeatedly feeding back the prediction error, the forecasts improve through time and converge to the observed values.

Chapter 7 described an approach that accounts for stream–aquifer interactions within the existing state–space structure of the model. The last chapters briefly discussed how the parameters of the model can be obtained. Parameter sensitivity was also mentioned. It turned out that the DLCM parameters, n and k, are stable, so they do not need to be continuously updated. Forecast accuracy of the coupled, deterministic–stochastic model was compared to a pure stochastic and the deterministic submodel part of the current model and it was shown that the coupled model performed the best, while the pure stochastic ARMA model performed the worst. Finally, illustrations of the Danube basin forecasting system were also provided.

Appendix I

A.I.1 STATE–SPACE DESCRIPTION OF LINEAR DYNAMIC SYSTEMS

The internal description of continuous, linear systems is given by the first-order ordinary differential equation

$$\dot{\mathbf{x}}(t) = \mathbf{F}(t)\mathbf{x}(t) + \mathbf{G}(t)\mathbf{u}(t) \tag{A1.1}$$

where $\mathbf{x}(t)$ is the n-dimensional *state variable*, $\mathbf{u}(t)$ is the p-dimensional *input variable*, $\mathbf{F}(t)$ is the $n \times n$ *state* or *system matrix*, and $\mathbf{G}(t)$ is the $n \times p$ *input matrix*. The dot denotes temporal differentiation. Eq. A1.1 describes the effect of inputs on the state of the system. The algebraic equation that relates the m-dimensional output, $\mathbf{y}(t)$, to the system state is

$$\mathbf{y}(t) = \mathbf{H}(t)\mathbf{x}(t) \tag{A1.2}$$

where $\mathbf{H}(t)$ is the $m \times n$ *output matrix*.

The continuous, linear dynamic system, described by the state (Eq. A1.1) and output equations (Eq. A1.2), is unambiguously characterized by the matrix-triplet

$$\mathbf{\Sigma}_C(t) = [\mathbf{F}(t), \mathbf{G}(t), \mathbf{H}(t)]$$

at each time-instant.

The solution (the equation of state-trajectory) of the state equation (e.g. Csáki, 1973) is given by

$$\mathbf{x}(t) = \mathbf{\Phi}(t, t_0)\mathbf{x}(t_0) + \int_{t_0}^{t} \mathbf{\Phi}(t, \tau)\mathbf{G}(\tau)\mathbf{u}(\tau)d\tau \tag{A1.3}$$

where $\mathbf{x}(t_0)$ is the initial state at time t_0 and $\mathbf{\Phi}(\cdot)$ is the $n \times n$ *state-transition matrix*. $\mathbf{\Phi}(\cdot)$ satisfies the following matrix differential equation

$$\frac{d}{dt}\mathbf{\Phi}(t, t_0) = \mathbf{F}(t)\mathbf{\Phi}(t, t_0) \tag{A1.4}$$

with initial condition

$$\boldsymbol{\Phi}(t_0, t_0) = \mathbf{I}_n$$

where $\mathbf{I}_n$ is the $n \times n$ identity matrix. With the help of the state-trajectory (Eq. A1.3), the output (Eq. A1.2) becomes

$$\mathbf{y}(t) = \mathbf{H}(t)\boldsymbol{\Phi}(t, t_0)\mathbf{x}(t_0) + \int_{t_0}^{t} \mathbf{H}(t)\boldsymbol{\Phi}(t, \tau)\mathbf{G}(\tau)\mathbf{u}(\tau)d\tau. \tag{A1.5}$$

In time-invariant systems the system-matrices are constant, i.e. $\boldsymbol{\Sigma}_C(t) = \boldsymbol{\Sigma}_C$, and the state-transition matrix depends only on the time elapsed: $\boldsymbol{\Phi}(t, t_0) = \boldsymbol{\Phi}(t - t_0)$. From Eq. A1.4 it follows that the state-transition matrix can be obtained as

$$\boldsymbol{\Phi}(t, t_0) = e^{(t-t_0)\mathbf{F}} \tag{A1.6}$$

which is the matrix-exponential of the system matrix. This way the output can be expressed as

$$\mathbf{y}(t) = \mathbf{H}e^{(t-t_0)\mathbf{F}}\mathbf{x}(t_0) + \int_{t_0}^{t} \mathbf{H}e^{(t-\tau)\mathbf{F}}\mathbf{G}\mathbf{u}(\tau)d\tau. \tag{A1.7}$$

If the system is relaxed initially, i.e. when $\mathbf{x}(t_0) = \mathbf{0}$, the output in Eq. A1.5 can be expressed as

$$\mathbf{y}(t) = \int_{t_0}^{t} \mathbf{\tilde{H}}(t, \tau)\mathbf{u}(\tau)d\tau \tag{A1.8}$$

where

$$\mathbf{\tilde{H}}(t, \tau) = \mathbf{H}(t)\boldsymbol{\Phi}(t, \tau)\mathbf{G}(\tau) \tag{A1.9}$$

is the *impulse–response matrix* of the system. In time-invariant systems $\mathbf{\tilde{H}}(t, \tau) = \mathbf{\tilde{H}}(t - \tau)$, by which Eq. A1.8 transforms into

$$\mathbf{y}(t) = \int_{t_0}^{t} \mathbf{\tilde{H}}(t - \tau)\mathbf{u}(\tau)d\tau \tag{A1.10}$$

which is the multi-variate form of *convolution*. With a choice of $t_0 = 0$

$$\mathbf{\tilde{H}}(t) = \mathbf{H}e^{t\mathbf{F}}\mathbf{G} \tag{A1.11}$$

can be written in a time-invariant case.

Eqs. A1.8 and A1.10 give an *external description* of linear dynamical systems.

So far, systems whose output did not depend explicitly on the input, only on the state of the system, were considered. When, however, the output is an explicit function of the input, the system is called *forward-coupled*. In such systems, only the output equation is changed; the state equation is the same, as before.

The output equation of a forward-coupled system is

$$\mathbf{y}(t) = \mathbf{H}(t)\mathbf{x}(t) + \mathbf{D}(t)\mathbf{u}(t) \tag{A1.12}$$

where $\mathbf{D}(t)$ is an $m \times p$ matrix. Assuming an initially relaxed system the output becomes

$$\mathbf{y}(t) = \int_{t_0}^{t} \mathbf{H}(t)\mathbf{\Phi}(t,\tau)\mathbf{G}(\tau)\mathbf{u}(\tau)d\tau + \mathbf{D}(t)\mathbf{u}(t)$$

which can be written with the help of the Dirac function as

$$\mathbf{y}(t) = \int_{t_0}^{t} [\mathbf{H}(t)\mathbf{\Phi}(t,\tau)\mathbf{G}(\tau) + \delta(t-\tau)\mathbf{D}(\tau)]\mathbf{u}(\tau)d\tau$$

which yields the impulse–response function of a forward-coupled system:

$$\mathbf{\hbar}(t,\tau) = \mathbf{H}(t)\mathbf{\Phi}(t,\tau)\mathbf{G}(\tau) + \delta(t-\tau)\mathbf{D}(\tau), \quad t \geqslant \tau. \tag{A1.13}$$

When the system is time-invariant this transforms into

$$\mathbf{\hbar}(t) = \mathbf{H}e^{t\mathbf{F}}\mathbf{G} + \delta(t)\mathbf{D}. \tag{A1.14}$$

In a linear dynamic system, all *structural properties* can be determined from analysis of the $\mathbf{\Sigma}_C$ matrix-triplet. Two such important properties are called *observability* and *controllability*.

Definition (Kalman): A linear, continuous, time-invariant dynamic system is observable, if $\mathbf{x}(t_0)$ can be determined from $\mathbf{u}(t)$ and $\mathbf{y}(t)$, $t_0 \leqslant t < \infty$. If this is true for any t_0, the system is *completely observable*.

Kalman also showed that a necessary and sufficient condition for a linear, continuous, time-invariant system to be observable is that

$$\left[\mathbf{H}^T \vdots \mathbf{F}^T\mathbf{H}^T \vdots (\mathbf{F}^T)^2\mathbf{H}^T \vdots \cdots \vdots (\mathbf{F}^T)^{n-1}\mathbf{H}^T \right] \tag{A1.15}$$

$n \times np$ hypermatrix have rank n, i.e. have n columns that are linearly independent. T denotes transpose.

Observability is a necessary condition for state-reconstruction and prediction. If a system is not observable, then its parameters cannot be identified.

For a discrete, linear, time-invariant system

$$\mathbf{x}_{t+1} = \mathbf{\Phi}\mathbf{x}_t + \mathbf{\Gamma}\mathbf{u}_t \qquad (A1.16)$$

$$\mathbf{y}_t = \mathbf{H}\mathbf{x}_t \qquad (A1.17)$$

the criterion for observability is similar (e.g. Csáki, 1973), namely, a necessary and sufficient condition for observability is that

$$\left[\mathbf{H}^T \vdots \mathbf{\Phi}^T\mathbf{H}^T \vdots (\mathbf{\Phi}^T)^2\mathbf{H}^T \vdots \cdots \vdots (\mathbf{\Phi}^T)^{n-1}\mathbf{H}^T\right] \qquad (A1.18)$$

$n \times np$ hypermatrix have rank n.

Observability requirements for a time-variant system can be found in Meditch (1969), where controllability properties, which we do not need for our forecasting, can also be found.

A.I.2   ALGORITHM OF THE DISCRETE LINEAR KALMAN FILTER

Let us assume that the discrete-time state equation (Eq. A1.16) contains an additive noise term

$$\mathbf{x}_{t+1} = \mathbf{\Phi}_{t+1,t}\mathbf{x}_t + \mathbf{\Gamma}_t\mathbf{u}_t + \mathbf{w}_t \qquad (A2.1)$$

where $\mathbf{x}_t$ is an n-dimensional state-variable, $\mathbf{\Phi}_{t+1,t}$ is an $n \times n$ state-transition matrix, $\mathbf{u}_t$ is a p-dimensional input, $\mathbf{\Gamma}_t$ is an $n \times p$ input-transition matrix, and $\mathbf{w}_t$ is an n-dimensional additive, stochastic sequence, and can be considered as model uncertainty, where we assume that it is a *Gaussian white noise* sequence with zero mean

$$E[\mathbf{w}_t] = \mathbf{0} \qquad (A2.2)$$

and covariance matrix

$$cov[\mathbf{w}] = E[\mathbf{w}_\tau\mathbf{w}_t^T] = \mathbf{Q}_t\delta_{\tau t} \qquad (A2.3)$$

where $\delta_{\tau t}$ is the Kronecker-delta symbol. Let's assume that the $n \times n$ $\mathbf{Q}_t$ matrix is positive semidefinite. Because of the above property of model uncertainty, the state variable is also a Gaussian stochastic variable, but it is not independent. Rather, due to Eq. A2.1, it is a *Markov sequence* with an initial mean value

$$E[\mathbf{x}_0] = \bar{\mathbf{x}}_0 \qquad (A2.4)$$

and initial $n \times n$ covariance matrix

$$cov[\mathbf{x}_0] = E[(\mathbf{x}_0 - \bar{\mathbf{x}}_0)(\mathbf{x}_0 - \bar{\mathbf{x}}_0)^T] = \mathbf{P}_0 \qquad (A2.5)$$

which is assumed to be positive semidefinite. Here it is also assumed that model uncertainty is independent of the initial state

$$E[(\mathbf{x}_0 - \bar{\mathbf{x}}_0)\mathbf{w}_t^T] = \mathbf{0}, \quad t \geqslant 0. \tag{A2.6}$$

As a consequence, the state variable, $\mathbf{x}_t$, is a *Gauss-Markov sequence*.

The input, $\mathbf{u}_t$, is deterministic; therefore it can be left out in the state and covariance estimation process. However, later it will be superimposed over the filtered variables during the calculation of their values. Thus, in deriving the filter-algorithm, the second term of the right-hand-side of Eq. A2.1 is neglected.

The output equation contains the m-dimensional output variable, $\mathbf{y}_t$

$$\mathbf{y}_t = \mathbf{H}_t \mathbf{x}_t \tag{A2.7}$$

where $\mathbf{H}_t$ is an $m \times n$ output matrix. Considering that the output measurements are laden with *measurement uncertainty*, $\mathbf{v}_t$, it is observed that

$$\mathbf{z}_t = \mathbf{y}_t + \mathbf{v}_t \tag{A2.8}$$

where $\mathbf{v}_t$ is assumed to be an additive, m-dimensional, Gaussian, white noise sequence with zero mean

$$E[\mathbf{v}_t] = \mathbf{0} \tag{A2.9}$$

and covariance matrix

$$cov[\mathbf{v}] = E[\mathbf{v}_\tau \mathbf{v}_t^T] = \mathbf{R}_t \delta_{\tau t}. \tag{A2.10}$$

Here $\mathbf{R}_t$ is assumed to be an $m \times m$ positive semidefinite matrix. With the help of Eq. A2.7, Eq. A2.8 can be written as

$$\mathbf{z}_t = \mathbf{H}_t \mathbf{x}_t + \mathbf{v}_t \tag{A2.11}$$

which is now the *measurement equation*. When all the state variables are measurable, the output matrix, $\mathbf{H}_t$, becomes the identity matrix.

Let's further assume that model and measurement uncertainties are independent of each other, i.e.

$$E[\mathbf{w}_\tau \mathbf{v}_t^T] = \mathbf{0}, \quad \forall(\tau, t). \tag{A2.12}$$

Eq. A2.11 generates a σ-algebra

$$\mathbf{Z}_t = [\mathbf{z}_1, \mathbf{z}_2, ..., \mathbf{z}_t] \tag{A2.13}$$

of the *measurement sequence* with the

$$\mathbf{Z}_t = [\mathbf{Z}_{t-1}, \mathbf{z}_t] \tag{A2.14}$$

chain-property.

Our objective is to specify the state variable, $\mathbf{x}_t$, from available measurements. Since we are dealing with probabilistic variables, this is an estimation problem.

The *estimation problem* is defined as: Given the measurement sequence in Eq. A2.13, an estimation of the state variable, $\mathbf{x}_t$, of the discrete dynamic system (described by Eq. A2.1) is sought which (a) is *unbiased*; (b) has *minimum variance*; and (c) is *consistent*.

The same problem can also be defined with a little more mathematical rigor as: Given the measurement sequence in Eq. A2.13, an unbiased estimation of the state variable is sought which minimizes the *loss function*, $L[\tilde{\mathbf{x}}\cdot]$, applied over the *estimation error*

$$\tilde{\mathbf{x}}_\tau = \hat{\mathbf{x}}_\tau - \mathbf{x}_\tau \tag{A2.15}$$

in conjunction with conditions specified in Eqs. A2.1 through A2.6, and A2.9 through A2.12. Since $\mathbf{x}_t$ is a probabilistic variable, so is $\hat{\mathbf{x}}_t$, and, thus, the loss function applied over it as well, having a minimum value in a statistical sense only. In the following, the expected value of the loss function will be referred to as the *expected loss*.

There are three types of the estimation problem, depending on the position of τ relative to t: (a) *filtering*, when $\tau = t$; (b) *smoothing*, when $\tau < t$; and (c) *forecasting*, when $\tau > t$. Because filtering is part of both the smoothing and forecasting problems, it will be discussed here in more detail, noting that forecasting becomes a simple task of matrix-manipulations once the filtered estimates have become available. The solution requires the following:

Thesis (Sherman, 1958): Let the $\mathbf{Z}_t$ measurement sequence and scalar-valued, convex, symmetric loss function, $L[\tilde{\mathbf{x}}\cdot]$, be given. The optimal estimation that minimizes the expected loss

$$E[L(\tilde{\mathbf{x}}\cdot)] \tag{A2.16}$$

is the conditional expectation

$$\hat{\mathbf{x}}_{\tau|t} = E[\mathbf{x}_\tau | \mathbf{Z}_t] \tag{A2.17}$$

where the $\mathbf{Z}_t$ condition is the measurement sequence, defined in Eq. A2.13.

The proof is simple, see e.g. Meditch (1969). As $\mathbf{x}_t$ is a Gauss-Markov sequence, it can be shown that its conditional value, with $\mathbf{Z}_t$ as condition, has an *n*-dimensional normal distribution (for each *t*) which is unambiguously characterized by its (time-varying) conditional expectation and covariance.

For solving the three estimation problems, these conditional statistics must be specified. It can be achieved in two ways: via either a direct or a recursive estimation approach. In real-time forecasting, it is practical to employ a recursive approach, since then the estimation procedure need not be performed repeatedly at each time-step when the latest measurements are incorporated into the sample. Rather, the "old" statistics, available prior to the latest measurements, can simply be modified (*updated*) with the latest data. Recursive estimation this way is a weighting of two uncertain pieces of information: the "old" estimation, which, by its very definition is laden with uncertainty; and the new measurements, which also contain uncertainties due to measurement errors (Eq. A2.11).

This way

[new estimation] = [old estimation] and [new measurements].

Kalman (1960) suggested a *linear combination* of these two uncertain pieces of information

$$\overset{\wedge}{\mathbf{x}}_{t|t} = \overset{\smile}{\mathbf{K}}_t \overset{\wedge}{\mathbf{x}}_{t|t-1} + \mathbf{K}_t \mathbf{z}_t \tag{A2.18}$$

where $\overset{\wedge}{\mathbf{x}}_{t|t-1}$ is the old, *a priori*, estimate of the conditional mean value of the state variable at time *t*, based on measurements, $\mathbf{Z}_{t-1}$, available up to time $(t-1)$, as condition; $\mathbf{z}_t$ are measurements obtained at time *t*; $\overset{\smile}{\mathbf{K}}_t$ and $\mathbf{K}_t$ are the two, yet unknown, weighting matrices; and $\overset{\wedge}{\mathbf{x}}_{t|t}$ is the new, *a posteriori*, estimate of the conditional mean value of the state variable at time *t*, using measurements, $\mathbf{Z}_t$, available up to time *t*, which now include the latest observations, $\mathbf{z}_t$, as condition. The objective is to obtain the weighting matrices.

Let's define the following estimation errors:

$$\overset{\smile}{\mathbf{x}}_{t|t} = \overset{\wedge}{\mathbf{x}}_{t|t} - \mathbf{x}_t \tag{A2.19}$$

which is called the *a posteriori error*, and

$$\overset{\smile}{\mathbf{x}}_{t|t-1} = \overset{\wedge}{\mathbf{x}}_{t|t-1} - \mathbf{x}_t \tag{A2.20}$$

which is the *a priori error*. Inserting Eq. A2.18 into Eq. A2.19 yields

$$\overset{\smile}{\mathbf{x}}_{t|t} = \overset{\smile}{\mathbf{K}}_t \overset{\wedge}{\mathbf{x}}_{t|t-1} + \mathbf{K}_t (\mathbf{H}_t \mathbf{x}_t + \mathbf{v}_t) - \mathbf{x}_t$$

where Eq. A2.11 was employed for z_t. Let's insert Eq. A2.20 into the above equation

$$\hat{\overset{\smile}{x}}_{t|t} = \overset{\smile}{K}_t(\hat{\overset{\smile}{x}}_{t|t-1} + x_t) + K_t(H_t x_t + v_t) - x_t$$

which, after rearrangement, yields

$$\hat{\overset{\smile}{x}}_{t|t} = \overset{\smile}{K}_t \hat{\overset{\smile}{x}}_{t|t-1} + K_t v_t + (\overset{\smile}{K}_t + K_t H_t - I) x_t. \tag{A2.21}$$

Let's assume that the *a priori* estimation error is unbiased

$$E[\hat{\overset{\smile}{x}}_{t|t-1}] = 0.$$

Because the measurement error, v_t, in Eq. A2.21 has already been assumed to have zero mean, the *a posteriori* error becomes unbiased, i.e.

$$E[\hat{\overset{\smile}{x}}_{t|t}] = 0$$

only, if the last term on the right-hand-side of Eq. A2.21 is zero, namely, when

$$\overset{\smile}{K}_t = I - K_t H_t. \tag{A2.22}$$

This equation relates the two weighting matrices. (Note that Eq. A2.21 is structurally the same as Eq. A2.18, with the only difference being that the estimation error now is updated by the measurement error.) With Eq. A2.22, the *a posteriori* estimation in Eq. A2.18 becomes

$$\hat{x}_{t|t} = (I - K_t H_t)\hat{x}_{t|t-1} + K_t z_t \tag{A2.23}$$

which after rearrangement yields

$$\hat{x}_{t|t} = \hat{x}_{t|t-1} + K_t(z_t - H_t \hat{x}_{t|t-1}). \tag{A2.24}$$

This equation specifies the extent of the prediction update, since the

$$v_t = z_t - H_t \hat{x}_{t|t-1} \tag{A2.25}$$

expression's second term is the *a priori* estimate of the new measurement by virtue of Eqs. A2.9 and A2.11, i.e.

$$\hat{z}_{t|t-1} = H_t \hat{x}_{t|t-1}. \tag{A2.26}$$

This way the $z_t - \hat{z}_{t|t-1}$ term in Eq. A2.24 represents the information the new measurement carries, and in doing so, the $K_t v_t$ term specifies the extent of the prediction update between the *a priori* and *a posteriori* estimation phases. When $\boldsymbol{v}_t = \boldsymbol{0}$, $\hat{\mathbf{x}}_{t|t} = \hat{\mathbf{x}}_{t|t-1}$, which shows that the *a posteriori* estimation is identical to the *a priori* estimation, because the new measurement, $\mathbf{z}_t$, did not contribute any useful information to the old one, used for the *a priori* estimation. $\boldsymbol{v}_t$ is called *innovation sequence*. It can be proven (Kailath, 1968), that the innovation sequence is a white noise for optimal estimations, which indicates that the information content of $\boldsymbol{v}_t$ is fully utilized in such cases. The initial value of the recursive state estimation algorithm, Eq. A2.24, is given by Eq. A2.4

$$\hat{\mathbf{x}}_{0|0} = \hat{\mathbf{x}}_0.$$

So far it has only been shown how the *a posteriori* conditional expectation of $\mathbf{x}_t$ can be obtained for unbiased estimates, i.e. when $E[\breve{\mathbf{x}}_{t|t}] = \boldsymbol{0}$. Next, the calculation of the *a posteriori* conditional covariance of the estimation error is discussed.

By definition, the covariance of the estimation error is

$$\mathbf{P}_{t|t} = E[\breve{\mathbf{x}}_{t|t}\breve{\mathbf{x}}_{t|t}^T] \tag{A2.27}$$

where $\mathbf{P}_{t|t}$ is an $n \times n$ covariance matrix. Inserting Eq. A2.21 into Eq. A2.27, and taking into consideration that the third term of its right-hand-side is zero, plus that $\breve{\mathbf{K}}_t$ is given by Eq. A2.22, yields

$$\mathbf{P}_{t|t} = E[(\mathbf{I} - \mathbf{K}_t\mathbf{H}_t)\breve{\mathbf{x}}_{t|t-1}\breve{\mathbf{x}}_{t|t-1}^T(\mathbf{I} - \mathbf{K}_t\mathbf{H}_t)^T] + E[\mathbf{K}_t\mathbf{v}_t\mathbf{v}_t^T\mathbf{K}_t^T] \tag{A2.28}$$

where the expectation of the cross-products between state and measurement error has vanished due to assumed independence of the two sequences

$$E[\breve{\mathbf{x}}_{\cdot}\mathbf{v}^T] = \boldsymbol{0}.$$

Applying Eq. A2.10 and defining the *a priori* covariance, similar to Eq. A2.27, the *a posteriori* conditional covariance can be expressed by the

$$\mathbf{P}_{t|t} = (\mathbf{I} - \mathbf{K}_t\mathbf{H}_t)\mathbf{P}_{t|t-1}(\mathbf{I} - \mathbf{K}_t\mathbf{H}_t)^T + \mathbf{K}_t\mathbf{R}_t\mathbf{K}_t^T \tag{A2.29}$$

recursive formula with the following initial value (Eq. A2.5)

$$\mathbf{P}_{0|0} = \mathbf{P}_0.$$

The $\mathbf{K}_t$ weighting matrix can be obtained in the following way. Let's define the expected loss in Eq. A2.16 as the expectation of a *quadratic form* involving estimation error

$$J = E[\widehat{\mathbf{x}}_{t|t}^{T} \mathbf{A} \widehat{\mathbf{x}}_{t|t}] \tag{A2.30}$$

where $\mathbf{A}$ is an arbitrary $n \times n$ semi-definite matrix. For simplicity let it be the identity matrix: $\mathbf{A} = \mathbf{I}$. The objective is to minimize the expected loss, which entails the unrestrained minimization of the estimation error's squared norm with respect to the $\mathbf{K}_t$ weighting matrix

$$\min_{\mathbf{K}_t}(J). \tag{A2.31}$$

Using the property of the scalar product, Eq. A2.30 can be written as

$$J = E\left[\sum_{i=1}^{n} \widehat{x}_{i,t|t}^{2}\right] \tag{A2.32}$$

which is the same as the sum of the *a posteriori* covariance matrix's elements in the main diagonal. This latter, by definition, is the *trace* (*Tr*) of the covariance matrix

$$J = Tr(\mathbf{P}_{t|t}). \tag{A2.33}$$

The optimal weighting matrix, $\mathbf{K}_t$, now results by the well-known differentiation rule

$$\frac{\partial}{\partial \mathbf{K}_t} Tr(\mathbf{P}_{t|t}) = \mathbf{0} \tag{A2.34}$$

although now with respect to a matrix.

Note 11.1: For a triple matrix product, the following identity is true

$$\frac{\partial}{\partial \mathbf{A}} Tr(\mathbf{ABA}^{T}) = 2\mathbf{AB}$$

provided, $\mathbf{B}$ is symmetric. The following is also true (e.g. Gertler, 1973)

$$\frac{\partial}{\partial \mathbf{A}} Tr(\mathbf{AC}) = \mathbf{C}^{T}.$$

Inserting Eq. A2.29 into Eq. A2.34 yields

$$-2(\mathbf{I} - \mathbf{K}_t\mathbf{H}_t)\mathbf{P}_{t|t-1}\mathbf{H}_t^{T} + 2\mathbf{K}_t\mathbf{R}_t = \mathbf{0}$$

which, after rearrangement, gives

$$\mathbf{K}_t = \mathbf{P}_{t|t-1}\mathbf{H}_t^{T}(\mathbf{H}_t\mathbf{P}_{t|t-1}\mathbf{H}_t^{T} + \mathbf{R}_t)^{-1}. \tag{A2.35}$$

The optimal weighting matrix, $\mathbf{K}_t$, is called the *Kalman matrix*, *Kalman gain*, or even *filter matrix*. By additional differentiation of Eq. A2.35, it can be shown that it indeed minimizes Eq. A2.33.

Let's now derive the *a priori* statistics, with consideration of the deterministic input. The estimation of the *a priori* conditional expectation requires taking the expected value of Eq. A2.1 with respect to the available $\mathbf{Z}_t$ measurement sequence, as a condition. Since the expected value of the model uncertainty, $\mathbf{w}_t$, is zero, so is its conditional expectation, from which it follows that

$$\hat{\mathbf{x}}_{t+1|t} = \mathbf{\Phi}_{t+1,t}\hat{\mathbf{x}}_{t|t} + \mathbf{\Gamma}_t\mathbf{u}_t \tag{A2.36}$$

which is a one-step conditional prediction. There remains the *a priori* conditional covariance of the estimation error to be specified. By definition it is

$$\mathbf{P}_{t+1|t} = E[\breve{\mathbf{x}}_{t+1|t}\breve{\mathbf{x}}_{t+1|t}^T]. \tag{A2.37}$$

With respect to Eqs. A2.36 and A2.1, the following can be written

$$\breve{\mathbf{x}}_{t+1|t} = \hat{\mathbf{x}}_{t+1|t} - \mathbf{x}_{t+1} = \mathbf{\Phi}_{t+1,t}\breve{\mathbf{x}}_{t|t} - \mathbf{w}_t$$

and so

$$\mathbf{P}_{t+1|t} = E[(\mathbf{\Phi}_{t+1,t}\breve{\mathbf{x}}_{t|t} - \mathbf{w}_t)(\mathbf{\Phi}_{t+1,t}\breve{\mathbf{x}}_{t|t} - \mathbf{w}_t)^T]$$

which yields, by considering Eqs. A2.3 and A2.27,

$$\mathbf{P}_{t+1|t} = \mathbf{\Phi}_{t+1,t}\mathbf{P}_{t|t}\mathbf{\Phi}_{t+1,t}^T + \mathbf{Q}_t. \tag{A2.38}$$

Here the assumed independence of the estimation and measurement errors, as well as the matrix product rule: $(\mathbf{AB})^T = \mathbf{B}^T\mathbf{A}^T$, were also exploited.

With the help of the Kalman matrix, Eq. A2.35, the *a posteriori* conditional covariance (Eq. A2.29) can be brought into a simpler form. For simplicity's sake, let's now disregard the time notation in the right-hand-side of Eq. A2.29, i.e.

$$\mathbf{P}_{t|t} = (\mathbf{I} - \mathbf{KH})\mathbf{P}(\mathbf{I} - \mathbf{KH})^T + \mathbf{KRK}^T \tag{A2.39}$$

and in Eq. A2.35, which is now written as

$$\mathbf{K}(\mathbf{HPH}^T + \mathbf{R}) = \mathbf{PH}^T. \tag{A2.40}$$

Rearranging Eq. A2.39, gives

$$\mathbf{P}_{t|t} = (\mathbf{P} - \mathbf{KHP})(\mathbf{I} - \mathbf{KH})^T + \mathbf{KRK}^T$$
$$= \mathbf{P} - \mathbf{KHP} - \mathbf{PH}^T\mathbf{K}^T + \mathbf{KHPH}^T\mathbf{K}^T + \mathbf{KRK}^T$$
$$= (\mathbf{I} - \mathbf{KH})\mathbf{P} - \mathbf{PH}^T\mathbf{K}^T + \mathbf{K}(\mathbf{HPH}^T + \mathbf{R})\mathbf{K}^T$$

which, due to Eq. A2.40 is

$$\mathbf{P}_{t|t} = (\mathbf{I} - \mathbf{KH})\mathbf{P} - \mathbf{PH}^T\mathbf{K}^T + \mathbf{PH}^T\mathbf{K}^T$$

and so

$$\mathbf{P}_{t|t} = (\mathbf{I} - \mathbf{K}_t\mathbf{H}_t)\mathbf{P}_{t|t-1} \tag{A2.41}$$

which is indeed much shorter than Eq. A2.29. By looking at the above formulae, a similarity to the recursive least squares (RLS) algorithm (Young, 1984) is obvious.

Finally, it can be concluded that the Kalman filter, as a recursive conditional state estimation algorithm, is in fact a sequence of *a priori* and *a posteriori* state estimations, which is an example of the *predictor–corrector* principle, shown in the following illustration:

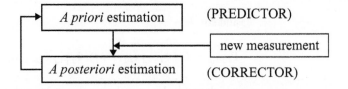

This also corresponds to the RLS principle. The two methods are practically the same in terms of estimation theory. The difference lies in the formulation of the problem and in the description of the system. A physically based state–space description is expected to incorporate more *a priori* information into the state-transition matrix than a purely statistical approach. Also, the Kalman filter algorithm incorporates measurement errors, while RLS does not. As a result, the Kalman filter gives superior estimates with noisy measurements when compared to RLS estimates, which explains the wide popularity of the Kalman filter algorithm (Szilágyi, 2004b).

The algorithm of the discrete linear Kalman filter is summarized below. See Gelb (1974), Meditch (1969), Sorenson (1966), and Young (1984) for further information on the algorithm and its generalizations.

As a final word on the Kalman filter, it should be noted that the Kalman gain, $\mathbf{K}_t$, can only contribute to the state estimation, if $\mathbf{R}_t$ is positive definite, in other words, if the measurements contain some uncertainty. When the measurements are considered error-free, the Kalman gain in

Eq. A2.35 degenerates into

$$\mathbf{K}_t = \mathbf{H}_t^{-1} \tag{A2.42}$$

and the *a posteriori* estimate in Eq. A2.24 transforms into

$$\hat{\mathbf{x}}_{t|t} = \mathbf{H}_t^{-1} \mathbf{z}_t \tag{A2.43}$$

while the corresponding *a posteriori* covariance, $\mathbf{P}_{t|t}$, becomes zero, and the *a priori* covariance, $\mathbf{P}_{t|t-1} = \mathbf{Q}_{t-1}$, i.e. equals model error covariance (Ahsan and O'Connor, 1994). Under such circumstances the Kalman - filter algorithm becomes identical to the RLS algorithm (Young, 1984).

The algorithm of the discrete linear Kalman filter:

$\mathbf{x}_{t+1} = \boldsymbol{\Phi}_{t+1,t}\mathbf{x}_t + \boldsymbol{\Gamma}_t\mathbf{u}_t + \mathbf{w}_t$ (State equation)

$\mathbf{z}_t = \mathbf{H}_t\mathbf{x}_t + \mathbf{v}_t$ (Measurement equation)

$\mathbf{w}_t \sim N(\mathbf{0}, \mathbf{Q}_t)$ (Noise statistics)

$\mathbf{v}_t \sim N(\mathbf{0}, \mathbf{R}_t)$

$E[\mathbf{x}_0] = \hat{\mathbf{x}}_0$ (Initial conditions)

$cov[\mathbf{x}_0] = \mathbf{P}_0$

$cov[\mathbf{x}_0, \mathbf{w}_t] = \mathbf{0}, \quad \forall t$

$cov[\mathbf{v}_\tau, \mathbf{w}_t] = \mathbf{0}, \quad \forall(\tau,t)$

$\hat{\mathbf{x}}_{t|t-1} = \boldsymbol{\Phi}_{t,t-1}\hat{\mathbf{x}}_{t-1|t-1} + \boldsymbol{\Gamma}_{t-1}\mathbf{u}_{t-1}$ (A priori state estimation)

$\mathbf{P}_{t|t-1} = \boldsymbol{\Phi}_{t,t-1}\mathbf{P}_{t-1|t-1}\boldsymbol{\Phi}_{t,t-1}^T + \mathbf{Q}_{t-1}$ (A priori state estimation)

$\mathbf{K}_t = \mathbf{P}_{t|t-1}\mathbf{H}_t^T(\mathbf{H}_t\mathbf{P}_{t|t-1}\mathbf{H}_t^T + \mathbf{R}_t)^{-1}$ (Weighting matrix)

New measurement : $\mathbf{z}_t$

$\hat{\mathbf{x}}_{t|t} = \hat{\mathbf{x}}_{t|t-1} + \mathbf{K}_t(\mathbf{z}_t - \mathbf{H}_t\hat{\mathbf{x}}_{t|t-1})$ (A posteriori state estimation)

$\mathbf{P}_{t|t} = (\mathbf{I} - \mathbf{K}_t\mathbf{H}_t)\mathbf{P}_{t|t-1}$ (A posteriori covariance estimation)

Appendix II

A.II.1 SAMPLE MATLAB SCRIPTS

1) fidemo.m

```
%State-transition matrix calculation

clear
n=3;k=.6;dt=1;
%Sample state-transition matrix (fi), Eq. 5-18
fi=zeros(n,n);
for i=1:n
   for j=1:i
      fi(i,j)=exp(-k*dt)*((k*dt)^(i-j))/prod(1:i-j);
   end
end
fi
```

Output:

```
fi =
0.5488  0       0
0.3293  0.5488  0
0.0988  0.3293  0.5488
```

2) gammademo.m

```
%Input-transition vector calculation

clear
n=3;k=.6;dt=1;
%Calculation of the input-transition vector of Eq. 5-22
gamv=zeros(n,1);
for i=1:n
   gamv(i)=(1/k)*gammainc(k*dt,i);
end
gamv
```

```
%Calculation of the input-transition vector of Eq. 6-8
gamv1=zeros(n,1);
for i=1:n
    gamv1(i)=(1/k)*gammainc(k*dt,i)*((-(k*dt)^(i-1) exp(-k*dt))/...
             (gammainc(k*dt,i)*gamma(i))+i/(k*dt));
end
gamv1

%Calculation of the input-transition vector of Eq. 6-9
gamv2=zeros(n,1);
for i=1:n
    gamv2(i)=(1/k)*gammainc(k*dt,i)*(1+((k*dt)^(i-1) *exp(-k*dt))/...
             (gammainc(k*dt,i *gamma(i))-i/(k*dt));
end
gamv2
```

Output:

```
gamv =
0.7520
0.2032
0.0385
gamv1 =
0.3386
0.1284
0.0280
gamv2 =
0.4134
0.0748
0.0105
```

3) PRdemo.m

```
%Pulse response calculation

clear
n=3;k=.6;dt=1;
H=zeros(1,n); %Output vector
H(n)=k; %The last element is k
%Sample state-transition matrix (fi), Eq. 5-18
fi=zeros(n,n);
for i=1:n
    for j=1:i
        fi(i,j)=exp(-k*dt)*((k*dt)^(i-j))/prod(1:i-j);
    end
end
```

```
%Calculation of the input-transition vector of Eq. 5-22
gamv=zeros(n,1);
for i=1:n
    gamv(i)=(1/k)*gammainc(k*dt,i);
end
%Calculation of the input-transition vector of Eq. 6-8
gamv1=zeros(n,1);
for i=1:n
    gamv1(i)=(1/k)*gammainc(k*dt,i)*((-(k*dt)^(i-1)*exp(-k*dt))/ ...
            (gammainc(k*dt,i)*gamma(i))+i/(k*dt));
end
%Calculation of the input-transition vector of Eq. 6-9
gamv2=zeros(n,1);
for i=1:n
    gamv2(i)=(1/k)*gammainc(k*dt,i)*(1+((k*dt)^(i-1) ...
            *exp(-k*dt))/(gammainc(k*dt,i) *gamma(i))-i/(k*dt));
end
for i=1:10 %The first 10 values
    UPR(i)=H*fi^(i-1)*gamv; %Unit-pulse response, Eq. 5-44
    DURR(i)=H*fi^(i-1)*gamv1; %Descending (from 1 to 0) unit-ramp
                            %response
    AURR(i)=H*fi^(i-1)*gamv2; %Ascending (from 0 to 1) unit-ramp
                            %response
end
The3PRs=[UPR' DURR' AURR']
```

Output:

```
The3PRs =
0.0231  0.0168  0.0063
0.0974  0.0547  0.0427
0.1489  0.0770  0.0719
0.1609  0.0801  0.0808
0.1465  0.0714  0.0751
0.1204  0.0579  0.0626
0.0925  0.0440  0.0485
0.0677  0.0320  0.0357
0.0478  0.0224  0.0253
0.0328  0.0153  0.0175
```

4) thetademo.m

%Observability matrix calculation

```
clear
n=3;k=.6;dt=1;
%Calculation of the state-transition matrix
```

```
fi=zeros(n,n);
for i=1:n
  for j=1:i
    fi(i,j)=exp(-k*dt)*((k*dt)^(i-j))/prod(1:i-j);
  end
end
%Calculation of observability (theta) matrix, Eq. 5-62
theta=zeros(n,n);
for i=1:n
  fin=fi^i;
  for j=1:n
    theta(i,j)=k*fin(n,j);
  end
end
theta
```

Output:

```
theta =
0.0593  0.1976  0.3293
0.1301  0.2169  0.1807
0.1607  0.1785  0.0992
```

5) dlcmdemo.m

```
%One-step forecast by the DLCM

clear %Clears the memory
clf %Erases the figure window

dt=1; %Time-step in days
k=1.2; %Storage coefficient [1/time]
n=2; %Number of storage elements
H=zeros(1,n); %Output vector
H(n)=k; %The last element is k

draw=1; %To have a plot: draw=1; not to: draw=0

%Concurrent daily in- (at Budapest) and output (Baja) discharge pairs
qin=[1084,1153,1580,3117,3575,3478,3324,3173,3042,2858,2741,...
    2553]';
qout=[1273,1286,1318,1536,2323,2985,3272,3230,3133,3025,2892,...
    2764]';
qinpred=qin; %Predictions for upstream station (LI-data framework),
             %in simulation mode the predicted inflows become
             %the observed ones
u1=qin(1:n); %Inflow at t
```

```
u2=qin(2:n+1); %Inflow at t+dt
y=qout(2:n+1); %Outflow at t+dt
pulse=0; %When 0 it is the LI-, when 1, pulse-data framework
%u2=u1;pulse=1; %With this, one can switch back to pulse data
                    %system
%%%%%%%%%%%%%%%%%%%%%%%%%%%%%%%%%%%%%%%%

%Calculation of the input transition vector of Eq. 6-8
gamv1=zeros(n,1);
for i=1:n
    gamv1(i)=(1/k)*gammainc(k*dt,i)*((-(k*dt)^(i-1)*exp(-k*dt))/...
            (gammainc(k*dt,i)*gamma(i))+i/(k*dt));
end

%Calculation of the input transition vector of Eq. 6-9
gamv2=zeros(n,1);
for i=1:n
    gamv2(i)=(1/k)*gammainc(k*dt,i)*(1+((k*dt)^(i-1)...
            *exp(-k*dt))/(gammainc(k*dt,i)*gamma(i))-i/(k*dt));
end


%%%%%%%%%%%%%%%%%%%%%%%%%%%%%%%%%%%%%%%%

%Calculation of the state-transition matrix (fi), Eq. 5-18
fi=zeros(n,n);
for i=1:n
  for j=1:i
    fi(i,j)=exp(-k*dt)*((k*dt)^(i-j))/prod(1:i-j);
  end
end


%%%%%%%%%%%%%%%%%%%%%%%%%%%%%%%%%%%%%%%%

%Calculation of observability (theta) matrix, Eq. 5-62
theta=zeros(n,n);
for i=1:n
  fin=fi^i;
  for j=1:n
    theta(i,j)=k*fin(n,j);
  end
end


%%%%%%%%%%%%%%%%%%%%%%%%%%%%%%%%%%%%%%%%

%Calculation of the pulse response functions (PRs)
%In the LI-data system there are two PRs
%One is the ascending unit ramp (from 0 to 1 over dt) response (AURR)
```

```
%The other is the descending unit ramp (from 1 to 0 over dt)
%response (DURR)
%When combined they yield the unit-pulse response (UPR) of
%the pulse-data system
h1=zeros(n); %DURR
h2=zeros(n); %AURR
tdmax=n;
h1=pr1(tdmax,n,k,dt); %Calls the function pr1
h2=pr2(tdmax,n,k,dt); %Calls the function pr2


%%%%%%%%%%%%%%%%%%%%%%%%%%%%%%%%%%%%%%%%%%%%

%Calculation of the e vector of Eqs. 5-68 and 6-20
e=zeros(n,1);
for i=1:n
    sumcum=0;
    for j=1:i
        sumcum=sumcum+h1(i-j+1)*u1(j)+h2(i-j+1)*u2(j);
    end
    e(i,1)=y(i)-sumcum;
end


%%%%%%%%%%%%%%%%%%%%%%%%%%%%%%%%%%%%%%%%%%%%

%Calculation of the initial state
xnull=inv(theta)*e; %Eqs. 5-69 & 6-20
%xnull=zeros(n,1); %Needed only when starting from a relaxed
                  %system
startday=1; %Startday of forecast error stats calc. (~5 for a relaxed
            %system)

%Here come the one-day forecasts%%%%%%%%%%%%%%%%%%%%%
x=xnull; %The state vector
for day=1:length(qout)-1
    if pulse==0
        x=fi*x+gamv2*qinpred(day+1)+gamv1*qin(day); %Eq. 6-7
    else
        x=fi*x+gamv2*qin(day)+gamv1*qin(day); %Eq. 5-15
    end
    yest(day,1)=H*x; %Eq. 5-14
end


%%%%%%%%%%%%%%%%%%%%%%%%%%%%%%%%%%%%%%%%%%%%

err=sum((yest(startday:end)-qout(1+startday:length(qout))).^2) ...
    /length(yest(startday:end)); %Mean-squared error (MSE)
```

%of forecasts

```
nsc=1-sum((yest(startday:end)-qout(1+startday:length(qout))) ...
    .^2)/sum((qout(1+startday:length(qout))-mean(qout+ ...
    startday-1)).^2); %Nash-Sutcliffe-type (NSC) forecast efficiency

if draw==1
  days=1:length(qin);
  plot(days,qout)
  hold on
  plot(days(2:end),yest,'rx')
  plot(days,qin,'g--') %The inflow
  legend('Measured stream-flow at Baja','1-day forecast', ...
         'Measured stream-flow at Budapest',2)
  xlabel('Days')
  ylabel('Q [m^{3}s^{-1}]')
end


%%%%%%%%%%%%%%%%%%%%%%%%%%%%%%%%%%%%%%%%%%%%%


MRSE=sqrt(err), nsc
yvsyest=[qout(2:length(qout)), yest] %observed & estimated pairs,
                                     %the first n predictions must equal
                                     %measured values
function h1=pr1(tdmax,n,k,dt) %Must be a separate file named pr1.m
for tdt=1:tdmax
    for i=1:n
        row(1,i)=((k*dt*(tdt-1))^(n-i))/prod(1:n-i);
        column(i,1)=gammainc(k*dt,i)*(i/(k*dt)-(k*dt)^(i-1) ...
                    *exp(-k*dt)/gammainc(k*dt,i)/prod(1:i-1));
    end
h1(tdt)=exp(-k*dt*(tdt-1))*row*column; %DURR
end


function h2=pr2(tdmax,n,k,dt) %Must be a separate file named pr2.m
for tdt=1:tdmax
    for i=1:n
        row(1,i)=((k*dt*(tdt-1))^(n-i))/prod(1:n-i);
        column(i,1)=gammainc(k*dt,i)*(1-(i/(k*dt)-(k*dt)^(i-1)...
                    *exp(-k*dt)/gammainc(k*dt,i)/prod(1:i-1)));
    end
h2(tdt)=exp(-k*dt*(tdt-1))*row*column; %AURR
end
```

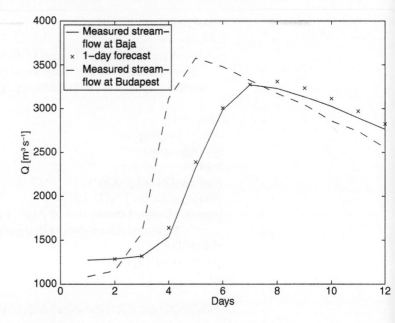

Output:

yvsyest =
1.0e+003 *
1.2860 1.2860
1.3180 1.3180
1.5360 1.6411
2.3230 2.3905
2.9850 3.0048
3.2720 3.2746
3.2300 3.3089
3.1330 3.2340
3.0250 3.1137
2.8920 2.9695
2.7640 2.8240

6) inputdetectiondemo.m

```
clear %Clears the memory
clf %Erases the figure window
dt=1; %Time-step in days
k=1.2; %Storage coefficient [1/time]
n=2; %Number of storage elements
H=zeros(1,n); %Output vector
H(n)=k; %The last element is k
%Concurrent daily in- (at Budapest) and output (Baja) discharge pairs
```

```
qin=[1084,1153,1580,3117,3575,3478,3324,3173,3042,2858, ...
     2741,2553]';
qout=[1273,1286,1318,1536,2323,2985,3272,3230,3133,3025, ...
     2892,2764]';
qinpred=qin; %Predictions for upstream station (LI-data framework),
             %in simulation mode the predicted inflows become
             %the observed ones
u1=qin(1:n); %Inflow at t
u2=qin(2:n+1); %Inflow at t+dt
y=qout(2:n+1); %Outflow at t+dt
%pulse=0; %When 0 it is the LI-, when 1, pulse-data framework
u2=u1;pulse=1; %With this, one can switch back to pulse data system

%%%%%%%%%%%%%%%%%%%%%%%%%%%%%%%%%%%%%%

%Calculation of the input-transition vector of Eq. 6-8
gamv1=zeros(n,1);
for i=1:n
    gamv1(i)=(1/k)*gammainc(k*dt,i)*((-(k*dt)^(i-1)*exp(-k*dt))/...
             (gammainc(k*dt,i)*gamma(i))+i/(k*dt));
end

%Calculation of the input-transition vector of Eq. 6-9
gamv2=zeros(n,1);
for i=1:n
    gamv2(i)=(1/k)*gammainc(k*dt,i)*(1+((k*dt)^(i-1) ...
             *exp(-k*dt))/(gammainc(k*dt,i) ...
             *gamma(i))-i/(k*dt));
end

%%%%%%%%%%%%%%%%%%%%%%%%%%%%%%%%%%%%%%

%Calculation of the state-transition matrix (fi), Eq. 5-18
fi=zeros(n,n);
for i=1:n
  for j=1:i
     fi(i,j)=exp(-k*dt)*((k*dt)^(i-j))/prod(1:i-j);
  end
end

%%%%%%%%%%%%%%%%%%%%%%%%%%%%%%%%%%%%%%

%Calculation of observability (theta) matrix, Eq. 5-62
theta=zeros(n,n);
for i=1:n
  fin=fi^i;
  for j=1:n
```

```
                    theta(i,j)=k*fin(n,j);
                end
            end


    %%%%%%%%%%%%%%%%%%%%%%%%%%%%%%%%%%%%%%%%%%%

    %Calculation of the pulse-response functions (PRs)
    %In the LI-data system there are two PRs
    %One is the ascending unit ramp (from 0 to 1 over dt) response (AURR)
    %The other is the descending unit ramp (from 1 to 0 over dt)
    %response (DURR)
    %When combined they yield the unit-pulse response (UPR) of
    %the pulse-data system
    h1=zeros(n); %DURR
    h2=zeros(n); %AURR
    tdmax=n;
    h1=pr1(tdmax,n,k,dt); %Calls the function pr1
    h2=pr2(tdmax,n,k,dt); %Calls the function pr2


    %%%%%%%%%%%%%%%%%%%%%%%%%%%%%%%%%%%%%%%%%%%

    %Calculation of the e vector of Eqs. 5-68 and 6-20
    e=zeros(n,1);
    for i=1:n
        sumcum=0;
        for j=1:i
            sumcum=sumcum+h1(i-j+1)*u1(j)+h2(i-j+1)*u2(j);
        end
        e(i,1)=y(i)-sumcum;
    end


    %%%%%%%%%%%%%%%%%%%%%%%%%%%%%%%%%%%%%%%%%%%

    %Calculation of the initial state
    xnull=inv(theta)*e; %Eqs. 5-69 & 6-20

    %Input detection starts here
    uest=zeros(length(qin),1);
    if pulse==0
      uest(1)=qin(1);
    else
      uest(end)=NaN; %The last (12th) inflow value cannot be estimated
                     %since that would require the 13th outflow value

    end
```

```
x=xnull; %The state vector
for day=2:length(qout)
   if pulse==0
      Eq. 6-23
      uest(day)=(1/(H*gamv2))*(qout(day)-H*fi*x-H*gamv1*uest...
            (day-1));
        x=fi*x+gamv2*uest(day)+gamv1*uest(day-1); %Eq. 6-24
   else
        uest(day-1)=(1/(H*(gamv1+gamv2)))*(qout(day)-H*fi*x);
                %Eq. 5-82
        x=fi*x+(gamv2+gamv1)*uest(day-1); %Eq. 5-83
   end
end


%%%%%%%%%%%%%%%%%%%%%%%%%%%%%%%%%%%%%%%%%%

days=1:length(qin);
plot(days,qin)
hold on
plot(days,uest,'rx')
legend('Observed stream-flow at Budapest', ...
'Detected stream-flow from observed values at Baja')
xlabel('Days')
ylabel('Q [m^{3}s^{-1}]')


%%%%%%%%%%%%%%%%%%%%%%%%%%%%%%%%%%%%%%%%%%

uest

Output:

uest =
1.0e+003 *
1.0840
1.1530
2.0294
3.5893
3.5070
3.4241
3.0023
3.0557
2.8736
2.7276
2.6219
NaN
```

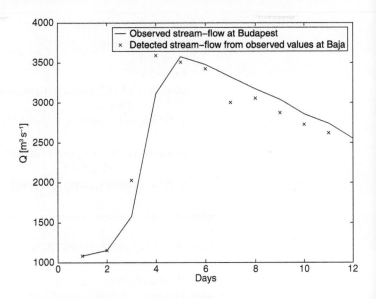

7) dlcmmultidemo.m

%Multi-step (1 through 3 days) forecast and
%parameter optimization with the
%DLCM. Optimization is achieved by
%a trial-and-error method of systematically
%changing (in two loops) the parameter (k & n)
%values of the cascade. The (k, n) set with the
%smallest simulation error is identified.
clear %Clears the memory
clf %Erases the figure window

%%

%Concurrent daily in- (at Budapest) and output (Baja) discharge pairs
qin=[1084,1153,1580,3117,3575,3478,3324,3173,3042,2858,2741,...
    2553]';
qout=[1273,1286,1318,1536,2323,2985,3272,3230,3133,3025,2892,...
    2764]';
qinpred=qin; %Predictions for upstream station (LI-data framework),
            %in simulation mode the predicted inflows become
            %the observed ones

%%

draw=1; %To have a plot: draw=1; not to: draw=0
pulse=0; %When 0 it is the LI-, when 1, pulse-data framework
errmin=10^20; %Initial forecast error for trial-and-error calibration

```
        %of k,n
dt=1; %Time-step in days

%%%%%%%%%%%%%%%%%%%%%%%%%%%%%%%%%%%%%

kstart=1.2;kstep=.1;kend=1.2; %These can be changed
nstart=2;nend=2; %These can be changed
for k=kstart:kstep:kend %Storage coefficient [1/time]
    for n=nstart:nend; %Number of storage elements

%%%%%%%%%%%%%%%%%%%%%%%%%%%%%%%%%%%%%

        u1=qin(1:n); %Inflow at t
        u2=qin(2:n+1); %Inflow at t+dt
        y=qout(2:n+1); %Outflow at t+dt
        % u2=u1;pulse=1; %With this, one can switch back to pulse
                        %data system
        H=zeros(1,n); %Output vector
        H(n)=k; %The last element is k

%%%%%%%%%%%%%%%%%%%%%%%%%%%%%%%%%%%%%

        %Calculation of the input-transition vector of Eq. 6-8
        gamv1=zeros(n,1);
        for i=1:n
            gamv1(i)=(1/k)*gammainc(k*dt,i)*((-(k*dt)^(i-1) ...
                    *exp(-k*dt))/(gammainc(k*dt,i)*gamma(i))+ ...
                    i/(k*dt));
        end

%%%%%%%%%%%%%%%%%%%%%%%%%%%%%%%%%%%%%

        %Calculation of the input-transition vector of Eq. 6-9
        gamv2=zeros(n,1);
        for i=1:n
            gamv2(i)=(1/k)*gammainc(k*dt,i)*(1+((k*dt)^(i-1) ...
                    *exp(-k*dt))/(gammainc(k*dt,i)*gamma(i))- ...
                    i/(k*dt));
        end

%%%%%%%%%%%%%%%%%%%%%%%%%%%%%%%%%%%%%

        %Calculation of the state-transition matrix (fi), Eq. 5-18
        fi=zeros(n,n);
        for i=1:n
            for j=1:i
                fi(i,j)=exp(-k*dt)*((k*dt)^(i-j))/prod(1:i-j);
            end
        end

%%%%%%%%%%%%%%%%%%%%%%%%%%%%%%%%%%%%%
```

```
%Calculation of observability (theta) matrix, Eq. 5-62
theta=zeros(n,n);
for i=1:n
   fin=fi^i;
   for j=1:n
      theta(i,j)=k*fin(n,j);
   end
end
```

%%%

```
%Calculation of the pulse-response functions (PRs)
%In the LI-data system there are two PRs
%One is the ascending unit ramp (from 0 to 1 over dt)
%response (AURR)
%The other is the descending unit ramp (from 1 to 0 over dt)
%response (DURR)
%When combined they yield the unit-pulse response (UPR) of
%the pulse-data system
h1=zeros(n); %DURR
h2=zeros(n); %AURR
tdmax=n;
h1=pr1(tdmax,n,k,dt); %Calls the function pr1
h2=pr2(tdmax,n,k,dt); %Calls the function pr2
```

%%%

```
%Calculation of the e vector of Eqs. 5-68 and 6-20
e=zeros(n,1);
for i=1:n
   sumcum=0;
   for j=1:i
      sumcum=sumcum+h1(i-j+1)*u1(j)+h2(i-j+1)*u2(j);
   end
   e(i,1)=y(i)-sumcum;
end
```

%%%

```
%Calculation of the initial state
 xnull=inv(theta)*e; %Eqs. 5-69 & 6-20
%xnull=zeros(n,1); %Needed only when starting from a
                   %relaxed system
```

%%%

```
startday=1; %Startday of forecast error stats calc. (~5 for
                 %a relaxed system)
tau=3; %Maximum forecast lead time, if 3 then 1,2, and
             %3 day forecasts are calc. If in a simulational
             %mode (i.e. future inflow is known, not estimated),
             %the multiple day forecasts become the
                 %one-day ones
err=zeros(tau,1); %Mean-squared error (MSE) of forecasts


%%%%%%%%%%%%%%%%%%%%%%%%%%%%%%%%%%%%%%%%

for i=1:tau
  x=xnull; %The state vector

  if i==1
    for day=1:length(qout)-tau %Equal # of forecasts
                                  %independent of lead time
      if pulse==0
        %Eq. 6-7
        x=fi*x+gamv2*qinpred(day+1)+gamv1*qin(day);
      else
        x=fi*x+gamv2*qin(day)+gamv1*qin(day); %Eq. 5-15
      end
      yest(day,i)=H*x; %Eq. 5-14
    end

    err(i)=sum((yest(startday:end,i)-qout(i+startday:length ...
        (qout)-tau+i)).^2)/length(yest(startday:end,i));
    %Nash-Sutcliffe-type (NSC) forecast efficiency
    nsc(i)=1-sum((yest(startday:end,i)-qout(i+startday: ...
        length (qout)-tau+i)).^2)/sum((qout(i+startday: ...
        length(qout)-tau+i)-mean(qout+startday-1)).^2);

    if draw==1
      days=1:length(qin);
      subplot(tau,1,i), plot(days,qout)
      hold on
      subplot(tau,1,i), plot(days(2:end-tau+i),yest(:,i),'rx')
      subplot(tau,1,i), plot(days,qin,'g--') %The inflow
      legend('Stream-flow at Baja','1-day forecast', ...
          'Stream-flow at Budapest',4)
    end
  else

    for day=1:length(qout)-tau %Equal # of forecasts
                                  %independent of lead time
```

```
                              sumcum=0;
                              qinest(1)=qin(day);
                              for jj=1:i %Recursive multiple-day forecast
                                            %calculations start
                                qinest(jj+1)=qinpred(day+jj);
                                fii=fi^(i-jj);
                                if pulse==0
                                  %LI-data system
                                  sumcum=sumcum+k*fii(n,:)*gamv2(:)*qinest ...
                                            (jj+1)+k*fii(n,:)*gamv1(:)*qinest(jj);
                                else
                                  %Pulse-data system
                                  sumcum=sumcum+k*fii(n,:)*gamv2(:)*qinest ...
                                            (jj)+k*fii(n,:)*gamv1(:)*qinest(jj);
                                end
                              end
                              fii=fi^i;
                              %Eqs. 5-41 (times H) & 6-21
                              yest(day,i)=H*fii*x+sumcum;
                              if pulse==0
                                %Eq. 6-7
                                x=fi*x+gamv2*qin(day+1)+gamv1*qin(day);
                              else
                                %Eq. 5-15
                                x=fi*x+gamv2*qin(day)+gamv1*qin(day);
                              end
                        end
                        err(i)=sum((yest(startday:end,i)-qout(i+startday:length ...
                              (qout)-tau+i)).^2)/length(yest(startday:end,i));
                        nsc(i)=1-sum((yest(startday:end,i)-qout(i+startday:length ..
                              (qout)-tau+i)).^2)/sum((qout(i+startday:length ...
                              (qout)-tau+i)-mean(qout+startday-1)).^2);
                        if draw==1
                          subplot(tau,1,i), plot(days,qout)
                          hold on
                          subplot(tau,1,i), plot(days(1+i:end-tau+i),yest(:,i),'rx')
                          if i==2
                            legend('Stream-flow at Baja','2-day forecast',4)
                            YLabel('Q [m^{3}s^{-1}]')
                          else
                            legend('Stream-flow at Baja','3-day forecast',4)
                            XLabel('Days')
                          end
                        end
                      end
                  end
            end

            %%%%%%%%%%%%%%%%%%%%%%%%%%%%%%%%%%%%%%%%%%%%%
```

```
        if mean(err)
          if min(xnull)
            %Optimized mean (of the different leadtimes) MSE
            errmin=mean(err);
            kopt=k; %Optimized k value
            nopt=n; %Optimized n value
            %Mean NSC of the different lead-time forecasts
            nsc=mean(nsc);
          end
        end


%%%%%%%%%%%%%%%%%%%%%%%%%%%%%%%%%%%%%%%%%%


    end
end

%Prints the calibrated k, n values and the MRSE
kopt,nopt
MRSE=sqrt(errmin)
%Observed & estimated pairs
yvsyest=[qout(2:length(qout)-tau+1), yest(:,1)]
%The first n predictions must equal measured values for a correct code

Output:

kopt =
1.2000
nopt =
2
MRSE =
71.0999
yvsyest =
1.0e+003 *
1.2860  1.2860
1.3180  1.3180
1.5360  1.6411
2.3230  2.3905
2.9850  3.0048
3.2720  3.2746
3.2300  3.3089
3.1330  3.2340
3.0250  3.1137
```

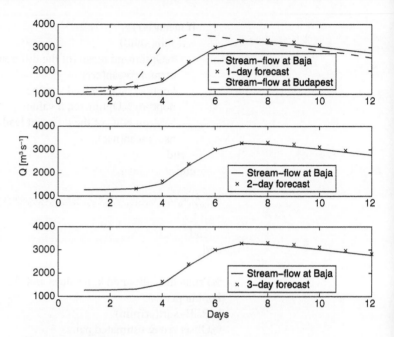

8) kalmandemo.m

%Demonstration of the Kalman filter for a) optimal
%predictions with noisy data; b) parameter
%estimation. The state equation now is scalar with no
%inputs (u), and the output matrix (H) unity. Both,
%model and measurement error, are prescribed as
%normally distributed noises with 0 means.
%Model parameter is also estimated by the
%Yule-Walker equation.

clear
clf

%%%

%These parameters can be modified by the user
n=1000; %Number of values to be generated
fi=.9; %Specified parameter of the AR(1) model
wstd=1; %Standard deviation of model error,
        %CANNOT BE ZERO!
vstd=1;   %Standard deviation of measurement error,
        %CANNOT BE ZERO!
Qcoef=1; %Since in reality model-error variance is
        %only estimated, Qcoef is an arbitrary
        %multiplier of true model variance. CANNOT BE ZERO!

Rcoef=1; %Same for measurement-error variance.
%CANNOT BE ZERO!
plotstart=1; %Starting value for plotting x and y
plotend=30; %Ending value for plotting x and y

%%%%%%%%%%%%%%%%%%%%%%%%%%%%%%%%%%%%%%%

toplot=plotstart:plotend;

w=normrnd(0,wstd,n,1);   %N(0,wstd) normally distr. number
%generation
v=normrnd(0,vstd,n,1); %N(0,vstd) normally distr. number generation

x(1)=0; %Initial value of the state
for i=2:n
   x(i)=fi*x(i-1)+w(i); %State eq., Eq. A2-1, with zero inputs [u(t)=0]
end
y=x+v'; %Measurement eq., Eq. A2-8, with H=1

ro1=[y(1:n-1); y(2:n)]';
r1=corrcoef(ro1);
%Estimation of the AR(1) parameter from the Yule-Walker eq.,
%Eq. 8-9
fiYW=r1(1,2)

yestYW=zeros(n,1);
yestYW(1)=mean(y); %The first predicted value is the mean of
%observations
for i=2:n
   yestYW(i)=fiYW*y(i-1); %One-step ahead prediction

end
subplot(2,1,1), plot(toplot,x(plotstart:plotend),'- -g')
xlabel('Selected period')
hold on
subplot(2,1,1), plot(toplot,y(plotstart:plotend))
subplot(2,1,1), plot(toplot,yestYW(plotstart:plotend),'ko')
%Mean-squared error (mse) of predictions
mseyYW=(yestYW'-y)*(yestYW'-y)'/(n-1);
%Mean-squared error (mse) of predictions related to x,
%which is typically unknown due to e.g., measurement error
msexYW=(yestYW'-x)*(yestYW'-x)'/(n-1);

%Sofar we assumed zero measurement error, i.e., y = x
%Below we account for the measurement error

```
%%%%%%%%%%%%%%%%%%%%%%%%%%%%%%%%%%%%%%%%%%%

Q=Qcoef*wstd*wstd;  %Estimation of model error (co)variance,
                    %Eq. A2-3
R=Rcoef*vstd*vstd; %Estimation of measurement error (co)variance,
                    %Eq. A2-10


%%%%%%%%%%%%%%%%%%%%%%%%%%%%%%%%%%%%%%%%%%%

xx=mean(y); %Estimation of the initial state, Eq. A2-4
xkalm(1)=xx; %Again, first prediction is just the mean of observations
mseyopt=1000000000;   %An arbitrarily large value for the
                        %optimization
Ktosee=zeros(n-1,1); %For plotting K
stdtosee=zeros(n-1,1); %For plotting sqrt(P)

for fiestopt=.5:.0001:1 %Loop for trial-and-error optimization of fi
    F=fiestopt;   %Here starts the Kalman-filter algorithm,
                    %see Appendix I
    P=var(y);   %Estimation of the initial state-prediction error
                %(co)variance which is equal to the initial state
                %(co)variance since the initial prediction is just the
                %mean
    xx=mean(y);
    Ktosee(1)=P/(P+R);
    stdtosee(1)=P;
    for i=2:n
        xx=F*xx; %A-priori state estimation
        xkalm(i)=xx;
        P=F*F*P+Q; %Estimate a-priori state-prediction error
                        %(co)variance
        K=P/(P+R); %Weighting factor (matrix) of Kalman
        Ktosee(i)=K;
        stdtosee(i)=sqrt(P);
        xx=xx+K*(y(i)-xx); %With the latest measur. update state estim.
        P=(1-K)*P; %Estimate a-posteriori state-prediction error (co)var.
    end
    mseytest=(xkalm-y)*(xkalm-y)'/(n-1);
    msextest=(xkalm-x)*(xkalm-x)'/(n-1);
    if mseytest
        mseyopt=mseytest; %Choosing the best AR(1) parameter estimate
        msexopt=msextest;
        fiopt=fiestopt;
        xkalmopt=xkalm;
```

```
        Kopt=Ktosee;
        stdopt=stdtosee;
    end
end

fiopt %This is the optimized AR(1) parameter

subplot(2,1,1), plot(toplot,xkalmopt(plotstart:plotend),'rx')
subplot(2,1,1), plot(toplot,xkalmopt(plotstart:plotend)+ ...
                    stdopt(plotstart:plotend)','b.')
legend('x','y','xestYW','xestKalman','Kalman-pred. error std')
subplot(2,1,1), plot(toplot,xkalmopt(plotstart:plotend)- ...
                    stdopt(plotstart:plotend)','b.')
subplot(2,1,2), plot(Kopt(1:10))
hold on
subplot(2,1,2), plot(stdopt(1:10),'r--')
xlabel('The first 10 values')
legend('K','P^{.5}')
mseratioy=mseyopt/mseyYW %Ratio of Kalman over
                           %Yule-Walker mse
                           %for y
mseratiox=msexopt/msexYW %Ratio of Kalman over
                           %Yule-Walker mse
                           %for x
```

Output:

```
fiYW =
0.7347
fiopt =
0.8905
mseratioy =
0.9346
mseratiox =
0.8720
```

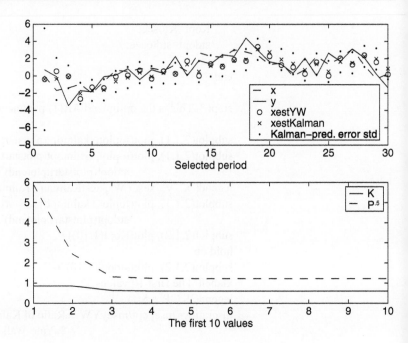

References

Abramowitz, M. and Stegun, I. A., *Handbook of Mathematical Functions*, Dover, New York, 1965.

Ahsan, M. and O'Connor, K. M., A reappraisal of the Kalman filtering technique, as applied in river flow forecasting, *J. Hydrol.*, 161, 197–226, 1994.

Ambrus, S. and Szöllősi-Nagy, A., *A Muskingum-Modell Általánosítása (Generalization of the Muskingum -Model*, in Hungarian), Lecture Notes, Budapest, 1984.

Ambrus, S., Borkert, M. and Ilse, J., Vizhozam előrejelzés a RIMO-modellel (Discharge forecasting with the RIMO model), *Vizügyi Közlemények*, 66(2), 219–229, 1984.

Andjelić, M. and Szöllősi-Nagy, A., On the use of stochastic structural models for real-time forecasting of river flow of the River Danube, *IAHS Publ.*, 129, 371–380, 1980.

Anderson, O. D., *Time Series Analysis and Forecasting: The Box-Jenkins Approach*, Butterworths, London, 1976.

Aoki, M., *Optimization of Stochastic Systems: Topics in Discrete Time Systems*, Academic Press, New York, 1967.

Bartha P., *Analógiás Berendezések Felhasználása az Árvizi Előrejelzésekben, Vizhozamelőrejelzések a Dunára (Application of analog devices for flood and flow forecasting of the Danube*, in Hungarian), VITUKI Report, 1970.

Bartha, P. and Szöllősi-Nagy, A., *A Tisza Vízjárásának Folyamatos Előrejel-zése a Fokozatosan Változó Nempermanens Vizmozgás Egyenleteinek Közelítő Megoldásával (Operational Forecasting Model for the Tisza River using Approximate Solutions of the Gradually Changing Non-Permanent Flow Equations*, in Hungarian), Research Report 721/1/12/1, VITUKI, Budapest, 1982.

Beck, M. B., *Real-Time Control of Water Quality and Quantity*, IIASA, RM-78-19, Laxenburg, 1978.

Becker, A. and Glos, E., Stufenmodell zur Hochwasserweltenberechnung in ausuferden Wasserläufen, *Wasserwirtschaft und Wassertechnik*, 20, 1970.

Box, G. E. P., Jenkins, G. M. and Reinsel, G. C., *Time Series Analysis, Forecasting and Control*, 3rd Edition, Prentice Hall International, 1994.

Bras, R. L. and Colón, R., Time-averaged mean of precipitation: estimation and network design, *Water Resour. Res.*, 14(5), 878–888, 1978.

Bras, T. L. and Rodriguez-Iturbe, I., *Random Functions and Hydrology*, Dover, New York, 1993.

Brebbia, C. A. and Ferrante, A., *Computational Hydraulics*, Butterworths, London, 1983.

Chiu, C.-L. and Isu, E. O., Stream temperature estimation using Kalman filter, *J. Hydr. Div. Proc. ASCE,* 104(HY9), 1,257–1,268, 1978.

Cooper, D. M. and Wood, E. F., Identification of multivariate time series and multivariate input–output models, *Water Resour. Res.*, 18(4), 937–946, 1982.

Csáki, F., *Fejezetek a Szabályozástechnikából: Állapotegyenletek (Chapters of Control Theory: State Equations*, in Hungarian), Műszaki Könyvkiadó, Budapest, 1973.

Cunge, J. A., On the subject of flood propagation computation method (Muskingum method), *J. Hydraul. Res.*, 7(2), 205–230, 1969.

Desoer, C. A., *Notes for a Second Course on Linear Systems*, van Nostrand, New York, 1970.

Dettinger, M. D. and Wilson, J. L., First order analysis of uncertainty in numerical models of groundwater flow, Part 1. Mathematical development, *Water Resour. Res.*, 17(1), 149–161, 1981.

Diskin, M. H. and Boneh, A., Properties of the kernels for time invariant, initially relaxed, second-order, surface runoff systems, *J. Hydrol.*, 17, 115–141, 1972.

Dooge, J. C. I., Linear theory of hydrologic systems, *USDA Tech. Bull.*, 1468, Washington DC, 1973.

Dooge, J. C. I. and O'Kane, J. P., *Deterministic Methods in System Hydrology*, A. A. Balkema, Lisse, 2003.

Dooge, J. C. I., Strupczewski, W. G. and Napiórkowski, J. J., Hydrodynamic derivation of storage parameters of the Muskingum method, *J. Hydrol.*, 54, 371–387, 1982.

Dooge, J. C. I., Kundzewicz, Z. W. and Napiórkowski, J. J., On backwater effects in linear diffusion flood routing, *Hydrol. Sci. J.*, 28(3), 391–402, 1983.

Duong, N., Winn, C. B. and Johnson, G. R., Modern control concepts in hydrology, *IEEE Trans. Systems, Man, Cybernetics*, SMC-5, 1, 46–53, 1975.

Dyhr-Nielsen, M., Loss of information by discretizing hydrologic series, *Hydrol. Papers*, No. 54, Fort Collins, 1972.

Eykhoff, P., *Systems Identification*, John Wiley, New York, 1975.

Fodor, G., *Lineáris Rendszerek Analízise (Analysis of Linear Systems*, in Hungarian), Műszaki Könyvkiadó, Budapest, 1967.

Forsythe, G. E. and Moler, C. B., *Computer Solution of Linear Algebraic Systems*, Prentice Hall, Englewood Cliffs, 1967.

Gabriel, K. R. and Neumann, I., A Markov chain model for daily runoff occurrence at Tel-Aviv, *Quart. J. Roy. Met. Soc.*, 88, 90–95, 1962.

Gelb, A. (ed.), *Applied Optimal Estimation*, MIT Press, Cambridge, 1974.

Georgakakos, K. P. and Bras, R. L., Real-time statistically linearized, adaptive flood routing, *Water Resour. Res.*, 18(3), 513–524, 1982.

Gertler, J., A Kálmán szűrő (The Kalman filter, in Hungarian), *Mérés és Automatika*, 21, 4, 122–125, 1973.

Hantush, M. M., Harada, M. and Marino, M. A., Hydraulics of stream flow routing with bank storage, *J. Hydrol. Engin.*, 7, 76–89, 2002.

Harkányi, K., Közvetlen optimalizáló eljárás hidrológiai előrejelző modellek paramétereinek gyors meghatározásához (Direct optimization for quick calculation of hydrologic model parameters, in Hungarian), *Vízügyi Közlemények*, 4, 1982.

Harkányi, K. and Bartha, P., Előrejelzési módszer kidolgozása a Tisza-Körösvölgyi rendszer üzemirányításához a vízfolyások belépő szelvényeire (Rainfall–runoff forecasting method for the management of the Tisza-Körösvölgyi water resources system), *VITUKI Research Report, #7623/1/171*, Budapest, 1984.

Hayami, S., *On the Propagation of Flood Waves*, Disaster Prevention Research Institute, Tokyo, 1951.

Henderson, F. M., Flood waves in prismatic channels, *J. Hyd. Div. Proc. ASCE*, 89, HY-1, 1979.

Hino, M. On-line prediction of hydrologic systems, *IAHR XVth Congress*, Istanbul, 1974.

Hostetter, G. H., Initial condition estimation in linear systems, *Proc. 15th ASILOMAR Conf. on Circuits, Systems and Computers*, 353–355, New York, 1982.

Hovsepian, K. K. and Nazarian, A. G., The solution of direct and inverse problems of outlet waves spreading on analogue computers, *Proc. IAHS/UNESCO Symp.*, Tucson, 1969.

Jones, S. B., Choice of space and time steps in the Muskingum–Cunge flood routing method, *Proc. Instn. Civ. Engrs.*, 71, 759–772, 1981.

Kailath, T., An innovation approach to least-square estimation, Part I: Linear filtering in additive white noise, *IEEE Trans. Automatic Control*, AC-13, 6, 646–655, 1968.

Kalinin, G. P. and Milyukov, P. I., Raschete neustanovivshegosya dvizheniya vody v otkrytykh ruslakh (On the computation of unsteady flow in open-channels, in Russian), *Met. i Gydrologia Zhurnal*, 10, 10–18, 1957.

Kalman, R. E., A new approach to linear filtering and prediction problems, *ASME J. Basic Eng.*, 82D, 35–45, 1960.

Kalman, R. E., On the general theory of control systems, *Proc. 1st IFAC Congress*, Moscow, 1961.

Kitanidis, P. K. and Bras, R. L., Real-time forecasting with a conceptual hydrologic model, *Water Resour. Res.*, 16(4), 1,025–1,044, 1980.

Kontur, I., A lefolyás álltalános lineáris modellje (Generalized linear model of runoff, in Hungarian), *Hidrológiai Közlöny*, 9, 404–412, 1977.

Koutitas, C. G., *Elements of Computational Hydraulics*, Pentech Press, London, 1983.

Kovács, G., A felszini lefolyás általános vizsgálata és az árvizek előrejelzése (Surface runoff investigations and flood forecasting, in Hungarian), *Vízügyi Közlemények*, 2, 199–252, 1974.

Kozák, M., *A Szabadfelszínű Nempermanens Vízmozgások Számítása Digitális Számítógépek Felhasználásával (Non-Permanent Open-Channel Flow Calculations on a Digital Computer,* in Hungarian), Akadémiai Kiadó, Budapest, 1977.

Kulandaiswamy, V. C., *A Basic Study of the Rainfall Excess Surface Runoff Relationship in a Basin*, Ph.D. Dissertation, University of Illinois, Urbana, 1964.

Kucsment, L. S., Solution of inverse problems for linear flow models, *Soviet Hydrol.*, 4, 1967.

Lighthill, M. J. and Whitham, G. B., On kinematic floods – flood movements in long rivers, *Proc. R. Soc. London*, A220, 281–316, 1955.

Mahmood, K. and Yevyevich, V., *Unsteady Flow in Open-Channels,* Water Resources Publications, Fort Collins, CO, 1975.

Maidment, D. R., *Stochastic State Variable Dynamic Programming for Water Resources Systems Analysis*, Ph.D. Dissertation, Univ. Illinois, Urbana, 1975.

Matalas, N. C., Statistics of a runoff-precipitation relation, *USGS Professional Papers*, 434-D, 1963.

McCarthy, G. T., The unit hydrograph and flood routing, Conf. North Atlantic Div., U.S. Corps of Engineers, New London, Conn., 1938.

Meditch, J. S., *Stochastic Optimal Linear Estimation and Control*, McGraw-Hill, New York, 1969.

Moler, C. B. and van Loan, D., Nineteen dubious ways to compute the exponential of a matrix, *SIAM Review*, 20, 801–835, 1978.

Moore, R. J. and Weiss, G., Real-time parameter estimation of a nonlinear catchment model using extended Kalman filters, in *Real-Time Forecasting/Control of Water Resource Systems* Wood and Szöllősi-Nagy (eds.), 1980.

Nash, J. E., The form of the instantaneous unit hydrograph, *Int. Assoc. Hydrol. Sci.*, 45(3–4), 114–121, 1957.

Nikolski, N. K., Operators, functions and systems: an easy reading, *Math. Surv. Monog. Amer. Math. Soc.*, No. 92–93, 2002.

O'Connell, P. E. and Clarke, R. T., Adaptive hydrological forecasting – a review, *Hydr. Sci. Bull.*, 25(6), 179–205, 1981.

O'Connor, K. M., Lineáris diszkrét hidrológiai kaszkád-modell (A linear, discrete, hydrologic cascade-model, in Hungarian), *VITUKI Tanulmányok és Kutatási Eredmények*, No. 47, Budapest, 1975.

O'Connor, K. M., A discrete linear cascade model for hydrology, *J. Hydrol.*, 29, 213–242, 1976.

Okunishi, K., Inverse transform of Duhamel integral for data processing in hydrology, *Bull. Disaster Prevent. Inst.*, 22(2), No. 201, Kyoto University, 1973.

Ponce, V. M., Linear reservoirs and numerical diffusion, *J. Hyd. Div., Proc. ASCE*, HY5, 691–699, 1980.

Prasad, R., A nonlinear hydrologic system response model, *Proc. ASCE J. Hyd. Div.*, 93, HY4, 201–221, 1967.

Press, W. H., Teukolsky, S. A., Vetterling, W. T. and Flannery, B. P., *Numerical Recipes*, University Press, Cambridge, 1986.

Price, R. K., Flood routing in natural rivers, *Proc. Inst. Civ. Eng.*, 55, 913–930, 1973.

Price, R. K., Flood routing studies, in *Flood Studies Report #3*, NERC, London, 1975.

Refsgaard, J. C. and Storm, B., MIKE SHE, in *Computer Models of Watershed Hydrology*, Singh, V. P. (ed.), Water Resources Publications, Highlands Ranch, CO, 1995.

Rényi, A., *Valószínűségszámítás (Probability Theory*, in Hungarian), Tan-könyvkiadó, Budapest, 1968.

Rózsa, P., *Lineáris Algebra (Linear Algebra*, in Hungarian), Műszaki Könyvkiadó, Budapest, 1974.

Sehitoglu, H., State estimation in linear dynamic systems via initial state identification, *Proc. 1982 American Control Conf.*, 2, 597–602, New York, 1982a.

Sehitoglu, H., On-line algorithms for initial state identification in linear systems, *Trans. ASME J. Dynamic System Measurement and Control*, 104, 114–117, 1982b.

Shaw, E. M., *Hydrology in Practice*, Van Nostrand Reinhold, London, 1983.

Sherman, S., Non-mean-square error criteria, *IRE Trans. Inform. Theory*, IT-4, 125, 1958.

Singh, V. P., *Kinematic Wave Modeling in Water Resources: Environmental Hydrology*, Wiley, New York, 1997.

Sorenson, H. W., Kalman filtering techniques, in *Advances in Control Systems*, Leondes, C. T. (ed.), Vol. 3, Academic Press, New York, 1966.

Stoker, J. J., *Numerical Solution of Flood Prediction and River Regulation Problems: Derivation of Basic Theory and Formulation of Numerical Methods of Attack*, Report I, IMM-NYU-200, New York University Institute of Mathematical Sciences, New York, 1953.

Strupczewski, W. and Kundzewicz, Z., Determination of structure and parameter of conceptual flood routing models, *Int. Conf. on Numerical Modeling of River, Channel and Overland Flow*, Pozsony, 1981.

Szigyártó, Z., A nempermanens viszonyok közötti vízhozam meghatározásának elméleti alapjai (Theoretical foundations of discharge calculations under nonpermanent flow conditions, in Hungarian), *Hidrológiai Közlöny*, 64(3), 148–153, 1984.

Szilágyi, J., Stage forecasting by an adaptive, stochastic model (in Hungarian), *Vízügyi Közlemények*, 74(1), 91–104, 1992.

Szilágyi, J., State–space discretization of the KMN-cascade in a sample-data system framework for streamflow forecasting, *J. Hydrol. Engin.*, 8(6), 339–347, 2003.

Szilágyi, J., Accounting for stream–aquifer interactions in the state–space discretization of the KMN-cascade for streamflow forecasting, *J. Hydrol. Engin.*, 9(2), 135–143, 2004a.

Szilágyi, J., Comment on "A reappraisal of the Kalman filtering technique, as applied in river flow forecasting" by M. Ahsan and K. M. O'Connor, *J. Hydrol.*, 285, 286–289, 2004b.

Szilágyi, J., Extension of the Discrete Linear Cascade Model for noninteger number of storage elements (in Hungarian), *Hidrológiai Közlöny*, 85(1), 37–41, 2005.

Szilágyi, J., Discrete state–space approximation of the continuous Kalinin–Milyukov–Nash cascade of noninteger storage elements, *J. Hydrol.*, 328, 132–140, 2006.

Szilágyi, J., Bálint, G., Gauzer, B. and Bartha, P., Flow routing with unknown rating curves using a state–space reservoir-cascade-type formulation, *J. Hydrol.*, 311, 219–229, 2005.

Szilágyi, J., Parlange, M. B. and Bálint, G., Assessing stream–aquifer interactions through inverse modeling of flow routing, *J. Hydrol.*, 327, 208–218, 2006.

Szilágyi, J., Pinter, N. and Venczel, R., Application of a routing model for detecting channel flow changes with minimal data, *J. Hydrol. Eng.*, 13(6), 521–526, 2008.

Szöllősi-Nagy, A., Hydrological and water resource systems from a state point of view: Estimation of state variables - the Kalman filtering, *Symposium on Mathematical Modelling in Hydrology*, Galway, 1974.

Szöllősi-Nagy, A., An adaptive identification and prediction algorithm for the real-time forecasting of hydrologic time-series, IIASA, RM-75-22, Laxenburg, 1975.

Szöllősi-Nagy, A., Szochasztikus irányítási modell vízfolyások oxigénháztartásának folyamatos szabályozásához (Stochastic process model for regulating stream oxigene balance, in Hungarian), *Hidrológiai Közlöny,* 10(11–12): 435–443, 523–530, 1979.

Szöllősi-Nagy, A., The discretization of the continuous linear cascade by means of state–space analysis, *J. Hydrol.*, 58, 223–236, 1982.

Szöllősi-Nagy, A., Input detection by the discrete linear cascade model, *J. Hydrol.*, 89, 353–370, 1987.

Szöllősi-Nagy, A. and Mekis, É., Primenenia filtru Kalmana dla korekcii prognozov po diskretnoi modeli karkadnovo tipa, Dunamenti Országok XII. Hidrológiai Előrejelzési Konferenciája (*12th Hydrological Forecasting Conference of the Danube Countries*), Bukarest, 1982.

Szöllősi-Nagy, A., Todini, E. and Wood, E. F., State–space model for real-time forecasting of hydrological time-series, *J. Hydrol. Sci.*, 4, 1–11, 1977.

Todini, E. and Bouillot, D., A rainfall–runoff Kalman filter model, IFIP Conference, Bruges, 1975.

Vágás, I., Önszabályozó általános rendszer láncolatok valószinűségi jellemzése (Probabilistic description of self-regulating generalized system sequences, in Hungarian), *Hidrológiai Közlöny*, 9, 1970.

Whitehead, P., Application of recursive estimation techniques to time-variable hydrological systems, *J. Hydrol.*, 40, 1–16, 1979.

Wood, E. F. and Szöllősi-Nagy, A., An adaptive algorithm for analyzing short-term structural and parameter changes in hydrologic prediction models, *Water Resour. Res.*, 14(4), 577–581, 1978.

Wood, E. F. and Szöllősi-Nagy, A. (eds.), *Real-Time Forecasting/Control of Water Resources Systems*, Pergamon Press, Oxford, 1980.

Woolhiser, D. A. and Liggett, J. A., Unsteady one-dimensional flow over a plane – The rising hydrograph, *Water Resour. Res.*, 3(3), 753–771, 1967.

Young, P. C., Recursive approaches to time-series analysis, *Bull. Inst. Maths. Appl.*, 10, 209–224, 1974.

Young, P. C., *Recursive Estimation and Time-Series Analysis*, Springer-Verlag, Berlin, 1984.

Young, P. C., Advances in real-time flood forecasting, *Phil. Trans. R. Soc. Lond.*, A(2002)360, 1,433–1,450, 2002.

Guide to the Exercises

CHAPTER 2

2.4. It must be shown that the integral is equal to unity.

CHAPTER 3

3.1. They represent the storage responses of the decreasing order cascades (starting with an order of n) to an input in the form of the Dirac-delta function. This way the impulse-response function of the n-cascade can be formulated in terms of storage and outflow, the latter being equal to the former multiplied by k. Note that because the state equation is written for storages, the impulse response of the system must originally be formulated for storage values. The impulse response of the cascade in terms of outflow results only via the output equation.

CHAPTER 4

4.1. It must be shown that the outflow of a single storage element (ke^{-tk}) when convoluted by itself yields the impulse response of the 2-cascade, i.e. $k^2 te^{-tk}$. Similarly, it can be shown, for example, that the output of the $(n-1)$-cascade when convoluted by (ke^{-tk}) yields the impulse response of the n-cascade.

4.2. It is easy to do the differentiation by hand for small values of n. For arbitrary n values try e.g. Maple or Mathematica.

CHAPTER 5

5.1. When $n = 1$, $i = 1$ in the definition of the incomplete gamma function. Therefore its integral form zero to $k\Delta t$ yields $1 - e^{-k\Delta t}$ which is the same as Eq. (5.20).

5.2. The response of the last storage element in a cascade is made up of the following individual responses: the response of the last storage element in the cascade, plus the response of the 2-cascade made up of the $(n-1)th$ and nth elements, plus the response of the 3-cascade made up of the $(n-2)nd$, $(n-1)th$ and nth elements, and so on. These responses of a relaxed system at time $t-0$ are obtained as the newly attained storage at time $t+0$ via an inflow in the shape of a Dirac-delta function into the first (and only the first) storage element within the cascade, multiplied by the impulse-response function of the relevant cascade.

5.3. It contains the unit-pulse responses due to the definition of the input signal, i.e. that it is constant over Δt.

5.4. The solution of $dx/dt = -kx$ with $x(0) = 1$ (since the inflow is in the form of a Dirac-delta function) becomes $x(t) = e^{-k(t-t_0)}$ which indeed satisfies Eq. (5.37).

5.6. It is $h_{i\Delta t} = e^{-(i-1)k\Delta t}(1 - e^{-k\Delta t})$.

5.7. For example, the convolution of the unit-pulse input with the impulse response function can be done in two steps. Up until Δt, the input is a constant, i.e. unity. At $t = \Delta t$ the output becomes $1 - e^{-k\Delta t}$, which is the unit-step response function. At $t = \Delta t$ the storage is $(1 - e^{-k\Delta t})/k$, so for $t \geq \Delta t$ the output is this storage multiplied by the impulse-response function, $ke^{-k(t-\Delta t)}$.

5.10. From Exercises 5.6 and 5.8 it follows.

5.12. $x_0 = 2420.1$. See Note 5.23.

5.13. $x_0 = [2050.7, 85.4]'$, $\hat{y}_3 = 1384.4$. Use Eq. 5.62 to obtain the observability matrix and then calculate its inverse. Use Eq. 5.70 for obtaining e_n in which the ordinates of the discrete unit-pulse response function can be obtained from Eq. 5.46 the easiest, making use of Eqs. 5.18 and 5.22. With the help of Eq. 5.41 the storages can be obtained for $t = 1, 2, 3$, step by step. The last element of the storage vector when multiplied by k yields the predicted outflow values at each time step. Note that the first two predictions are perfect (i.e. they equal the observed values up to some rounding errors) if you did the calculations correctly. This is not surprising since these two outflow values were known for calculating the initial state x_0 for $n = 2$.

CHAPTER 6

6.1. Let's consider the linear change from a to b over Δt as depicted below.

We want to show that at time $t + c$ segments '=' plus d indeed equal the value the linear signal assumes at $t+c$. For that we simply need to show that the two segments marked by '=' are equal. This can be seen by considering that $\tan(\beta) = (1+x)\tan(\alpha)$. Using the definition of the tangent yields $d/c = (1+x)e/c$, i.e. $d = (1+x)e$. But then it follows immediately that the two segments marked by '='

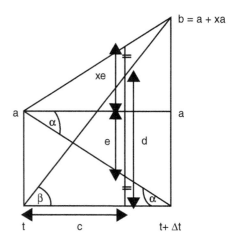

are indeed equal. For a decreasing signal we obtain the same situation by defining c as the time needed to reach $t + \Delta t$. This concludes the proof.

6.4. For a Dirac-delta input the output of the continuous 2-cascade (k_1 and k_2) is $\int_0^t k_1 e^{-k_1 \tau} k_2 e^{-k_2(t-\tau)} d\tau$, while the same for the rearranged cascade becomes $\int_0^t k_2 e^{-k_2 \tau} k_1 e^{-k_1(t-\tau)} d\tau$, which we know is the same as before since τ and $(t-\tau)$ are interchangeable within the convolution integral.

6.5. $x_0 = 2368.1$. See 6.6 for an explanation.

6.6. $x_0 = [1524.7, 690.5]'$, $\widehat{y}_3 = 1641.1$. Use Eq. 5.62 to obtain the observability matrix and then calculate its inverse. Use Eqs. 5.20, 5.21, plus 6.16 through 6.20 for obtaining e_n, making use of Eqs. 6.8 and 6.9. With the help of Eq. 6.7 the storages can be obtained for $t = 1, 2, 3$, step by step. The last element of the storage vector when multiplied by k yields the predicted outflow values at each time step. Note that the first two predictions are perfect (i.e. they equal the observed values up to some rounding errors) if you did the calculations correctly.

CHAPTER 8

8.1. Naturally, the Kalman filter results in better forecasts and yields very accurate estimate of the prescribed AR(1) parameter, while the Yule-Walker equation gives an erroneous parameter estimate whenever a measurement error is present. In the absence of the latter the two methods give identical parameter estimates and forecasts.

Subject index

For Product Safety Concerns and Information please contact our EU
representative GPSR@taylorandfrancis.com Taylor & Francis Verlag GmbH,
Kaufingerstraße 24, 80331 München, Germany

Printed and bound by CPI Group (UK) Ltd, Croydon, CR0 4YY
01/05/2025
01858476-0001